The Arlington Practical Botany/Book 5
PLANT PHYSIOLOGY

The Arlington Practical Botany

General Editor: Mary-Anne Burns, B.Sc. Hons (Lond), M.I.Biol.

PUBLISHED

Book 1/ *Plant Anatomy*

Book 5/ *Plant Physiology*

IN PREPARATION

Book 2/ *Algae, Fungi, Lichens and Bacteria*

Book 3/ *Bryophyta and Pteridophyta*

Book 4/ *Gymnospermae and Angiospermae*

Book 6/ *This volume will deal with the remaining botanical topics not covered by Books 1-5*

The Arlington Practical Botany/Book 5

PLANT PHYSIOLOGY

by MARY-ANNE BURNS *B.Sc. Hons (Lond), M.I.Biol.*
Formerly Lecturer at Brighton Technical College
and
ROY HUGHES, *B.Sc. (Hons), M.I.Biol.*
Lecturer in the Department of Applied Biology
at Brighton Technical College

London ARLINGTON BOOKS LTD.

PLANT PHYSIOLOGY

first published 1972 *by Arlington Books Ltd.*

15 *Duke Street, St James's, London, S.W.*1.

© *Mary-Anne Burns and Roy Hughes* 1972

printed by Martins of Berwick

SBN 85140-129-5

Contents

Plant and water relationships 1
    Osmosis 1
    Transpiration 9
    Osmosis Experiments 13
    Transpiration Experiments 20

Photosynthesis 31
    Photosynthesis experiments 37

Respiration 44
    Respiration experiments 53

Biochemistry 58
    The chemistry of plant products 58
    Biochemical tests 66
    Enzymes 71
    Enzyme experiments 75

Chromatography 81
    Chromatography experiments 89

Growth 97
    Growth experiments 103

Plant Movement 111
    Tropisms 111
    Nastisms 114
    Tactic movements 117
    Hygroscopic movements 117
    Turgur movements 119
    Autonomic movements 119
    Experiments to demonstrate Tropic responses 120
    Experiments to demonstrate Autonomic movements 125

Soil 126
    Soil study experiments 132

Subject Index 138

Index of Experiments 140

Editor's Note

WHILST STUDYING BOTANY for various examinations, and later when demonstrating at Brighton Technical College, I became aware of the lack of practical reference books, and in this series of six books I am attempting to fill this need. They are intended for students studying for Advanced and Scholarship level G.C.E. and similar examinations, but they will also prove valuable to those studying privately or at evening classes where time for practical work is limited.

Plant Physiology, the second book to be published, is designed so that students can follow the experiments and the theories behind them without having to consult other reference books at the time. Each section commences with explanatory text which introduces the student to the subject and, at the same time, covers all the points required for a full understanding of the subject, while dealing with the more important points in some detail.

The explanatory text is followed by a series of experiments, some of which quite simply demonstrate the theories mentioned in the text while others are designed to make the student think and work things out for himself. Diagrams of apparatus, etc., are included where they augment the information given in the text or experiments.

I would like to express my gratitude to Dr. Clifford Mortlock, Principal of the Jersey College of Further Education, St. Helier, for his great help in reading the proofs; and to Mr. George Martin, Junior, the printer, for his advice and co-operation throughout the production of the book.

Mary-Anne Burns
December, 1971

PLANT AND WATER RELATIONSHIPS

Water plays an important part in the life of the plant. It is the most abundant component of living cells constituting over 90% of the fresh weight of some tissues. The absence of adequate water supplies results in a drop in the turgidity of the living plant cells; these play an important mechanical role in herbaceous plants and a reduction in their turgidity results in wilting in these plants. The presence of water is essential for the germination of seeds, for photosynthesis and for respiration, and shortage of water is often the limiting factor governing plant growth.

Plants obtain the water they require from the soil in which they are rooted, the water entering through the root hairs of the piliferous region of the roots. Plants are also continually losing water to the air surrounding them through the process of transpiration (see page 9). Plants living in an environment where water is either in short supply or in a form in which it cannot readily be absorbed by the plant are modified to reduce the amount of water lost by transpiration. The various modifications are dealt with on page 12 (also in Book 1, page 88).

The various factors governing the entry of water into, and its passage through the plant are dealt with on pages 6 and 7. The process of transpiration is dealt with on page 9.
The movement of water between living cells is governed by the phenomenon of osmosis.

OSMOSIS

Osmosis is essentially a process of diffusion. It is a term used with reference to the passage of the solvent from a solution of low concentration to a similar solution of higher concentration through a membrane which is permeable to the solvent but not to the solute.

Consider first a solution such as sugar (solute) in water (solvent), separated from pure water by a membrane which is permeable to both solvent and solute, (diagram 1). According to the laws governing the movement of molecules in a liquid (the Kinetic Theory), the molecules of both solute and solvent are continually moving at random within the liquid and are continually striking and, since it is permeable, passing through the membrane in either direction. At first only water passes from B to A but both sugar and water pass in the other direction from A to B. As the concentration of sugar molecules in B increases so some will move back to A. Thus, at first the number of water molecules diffusing from B to A was greater than the number diffusing from A to B and there is a net increase in the number of water molecules in A. The water is said to have a **partial pressure** or **diffusion pressure** causing it to move from one compartment to the other. At first the diffusion pressure of water in B is greater than the diffusion pressure of water in A and it can be said that there is a diffusion pressure deficit of water in compartment A, which results in a net increase in the number of water molecules in A. Gradually an equilibrium is established between the two compartments as the number of water and sugar molecules passing in either direction evens out by which time the two solutions will be of equal concentration.

Diagram 1

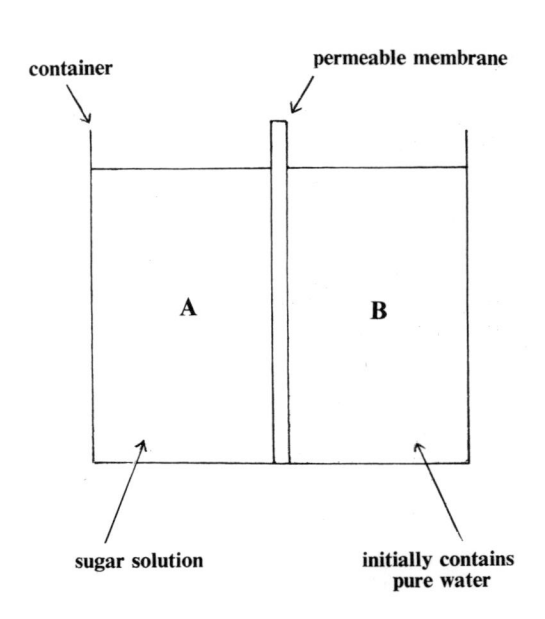

container — permeable membrane

A | B

sugar solution | initially contains pure water

Diagram 2

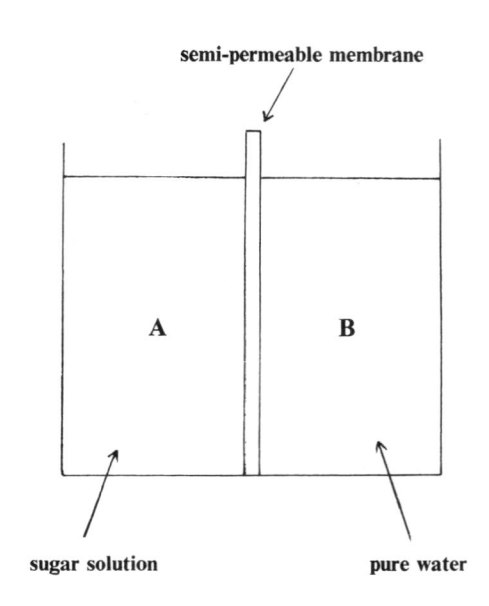

If the membrane instead of being permeable to both the solvent and the solute is permeable only to the solvent (a semi-permeable membrane) then water molecules pass in either direction as before with the resulting net increase in the number of water molecules in compartment A due to the diffusion pressure deficit of water in A. The passage of sugar molecules from A to B is prevented, since the membrane is impermeable to them. Provided it is unhindered the accumulation of water molecules in A will continue until the sugar solution reaches infinite dilution.

This passage of a solvent through a membrane which is permeable only to the solvent, from a region in which the solvent has a high diffusion pressure (dilute solution) to a region in which the solvent has a low diffusion pressure (concentrated solution), in an attempt to equalise the diffusion pressure deficit is referred to as **osmosis.**

Pure water has maximum diffusion pressure.

OSMOTIC PRESSURE

Diagram 3

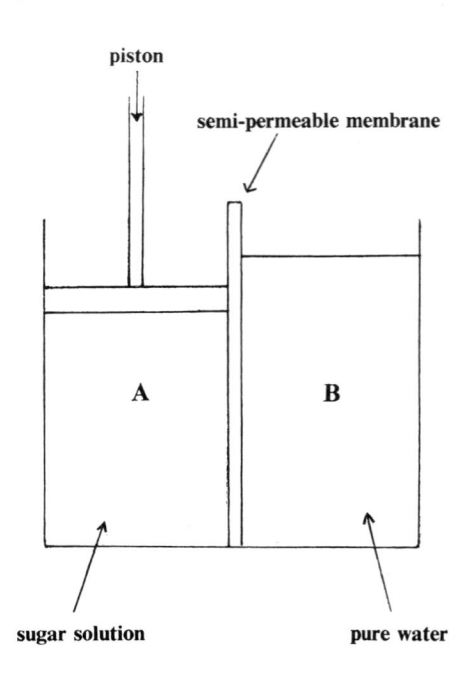

If the liquid in compartment A is completely enclosed, one side of the compartment being operable as a piston (diagram 3), then as water passes from B to A there is an increase in the volume of the liquid in compartment A and the piston is pushed out. It is possible to prevent this increase in volume and to keep the position of the piston constant by applying a pressure to the piston. The application of pressure to resist the increase in volume results in an increase in the kinetic energy of the molecules in A and although there are less water molecules in A they will move faster and strike the membrane more often than those in B, with the result that, in a given time, equal numbers will strike and pass through in either direction and no net movement of water (solvent) takes place. The pressure applied to a solution in this way, which is just sufficient to prevent the passage of the solvent into it is referred to as the **osmotic pressure** of the solution. The osmotic pressure is dependent on the concentration of the solution, a high concentration of solute means a high osmotic pressure; pure water has an osmotic pressure of zero.

It should be noted that the diffusion pressure and osmotic pressure are not the same.

For a simple laboratory experiment to demonstrate osmotic pressure see page 13.

Solutions having the same osmotic pressure are said to be **isotonic.** If there are two solutions with different osmotic pressures, then the more concentrated solution is said to be **hypertonic,** in relation to the weaker solution which is referred to as **hypotonic.**

THE WATER RELATIONSHIPS OF A LIVING VACUOLATED PLANT CELL

Considering the plant cell as part of an osmotic system

A living plant cell consists of a living protoplast surrounded by a cell wall which is secreted by the protoplast. The protoplast is composed of various protoplasmic and non-protoplasmic components including the nucleus and one or more vacuoles which are filled with the fluid cell sap. This cell sap consists of a solution in water of various salts and organic substances and would be capable, under the right conditions, of acting as an osmotic solution and it may therefore be said to have osmotic potential.

The cell wall is completely permeable to water and solutes which can therefore pass through it quite freely in either direction. The cytoplasm is permeable to water, and to a very limited number of solutes, some of which can pass through it quite rapidly, and it could therefore, in certain circumstances, act as a semi-permeable membrane since it surrounds the vacuoles which are filled with the fluid cell sap.

The cytoplasm has an outer limiting membrane the plasmalemma, and an inner membrane, the tonoplast, and it is at present not clear whether it is the plasmalemma, the tonoplast or the whole cytoplasm which is the actual semi-permeable membrane.

Thus an isolated vacuolated plant cell placed in water would form an osmotic system (one in which osmosis can take place). The cell must not be killed during the isolation for the dead protoplast does not function as a semi-permeable membrane.

Diagrammatic representation of a living plant cell

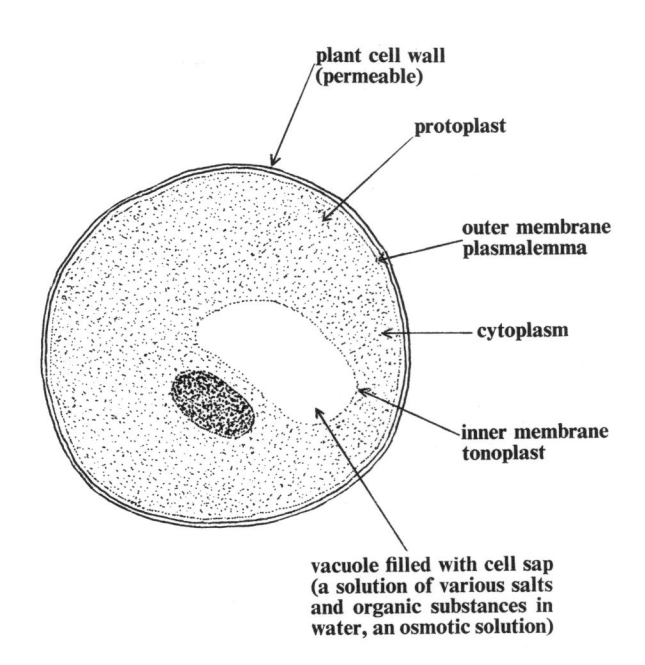

- plant cell wall (permeable)
- protoplast
- outer membrane plasmalemma
- cytoplasm
- inner membrane tonoplast
- vacuole filled with cell sap (a solution of various salts and organic substances in water, an osmotic solution)

Considering an isolated vacuolated living plant cell immersed in water

If an isolated living vacuolated plant cell is placed in water then an osmotic system exists. The cytoplasm acts as a semi-permeable membrane between the osmotically active cell sap and the external water. Water passes through the permeable cell wall, then through the cytoplasm to the vacuole, because the diffusion pressure of water in the vacuole is less than the diffusion pressure of the external water. Water continues to enter the vacuole and as the contents increase in size so the cell wall, which is elastic, stretches; as it stretches so it exerts an opposite and inwardly directed pressure on the contents. This pressure is referred to as the **wall pressure**. The name given to the internal hydrostatic pressure exerted by the contents on the wall is the **turgor pressure**. The wall pressure tends to oppose the entry of water into the cell and a point is eventually reached at which the wall pressure is so great that it prevents the entry of any more water although an osmotic equilibrium between the cell sap and the water outside has not been reached.

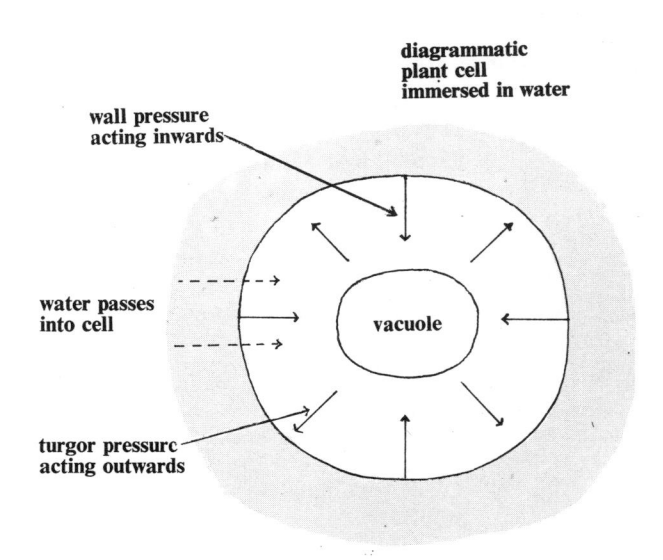

diagrammatic plant cell immersed in water

- wall pressure acting inwards
- water passes into cell
- vacuole
- turgor pressure acting outwards

The factors concerned in osmosis in the living vacuolated plant cell may be related by equations in which:—
- P_1 is the osmotic pressure of the cell sap
- P_2 is the osmotic pressure of the external solution (in this case water which is zero)
- W is the wall pressure
- S is a measure of the water absorbing capacity of the cell; in the past this has been termed the **suction pressure** of the cell but there is a tendency nowadays for this to be discarded for the term **diffusion pressure deficit**

$$S = (P_1 - P_2) - W \qquad\qquad S = P_1 - W$$

net suction pressure $\qquad\qquad\qquad\qquad\qquad$ gross suction pressure

The suction pressure as expressed in the first equation is the **net suction pressure** and takes into account the osmotic pressure of the external solution. The suction pressure can however be expressed as the **gross suction pressure,** this being the suction pressure of the cell when it is placed in pure water and is equal to the difference between the osmotic pressure of the cell sap and the wall pressure of the cell, as in the second equation.

If $S = P_2$ then no net movement of water will take place.

An alternative term the **water diffusion potential** abbreviated to **water potential** may also be used. This measures the tendency for water to diffuse outwards from the cell if the cell is placed in pure water. Since water will actually move inwards the water potential of a cell is usually a negative quantity. The value is the same as that of the diffusion pressure deficit with the sign reversed. A cell with a DPD of 15 atmospheres will have a water potential of -15 atmospheres.

When a cell guttates water the water potential has a positive value.

Considering the condition of an isolated living vacuolated plant cell placed in water

1. CONSIDERING FIRST A FLACCID CELL IN WHICH THE CONTENTS REST LIGHTLY AGAINST THE WALL

Here the wall pressure is zero; the osmotic pressure of the cell sap P_1 is much greater than the osmotic pressure of the external solution which being water is zero, thus there is a diffusion pressure deficit of water in the vacuole and the water absorbing power of the cell is equal to the difference in the osmotic pressures. Since the osmotic pressure of water is zero the water absorbing power of the cell is equal to its osmotic pressure.

$$W = 0 \text{ and } P_2 = 0$$
$$S = (P_1 - 0) - 0$$
$$S = P_1$$

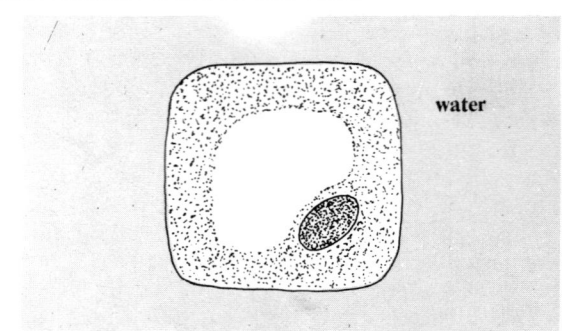

2. CONSIDERING A CELL WHICH IS FULLY TURGID

In this case the wall pressure is at its maximum. The osmotic pressure of the cell sap is less than that of the sap in the flaccid cell but it is still greater than that of water. However when the wall pressure is maximum the suction pressure is zero since the wall pressure opposing the entry of water is equal to the osmotic pressure tending to draw the water in.

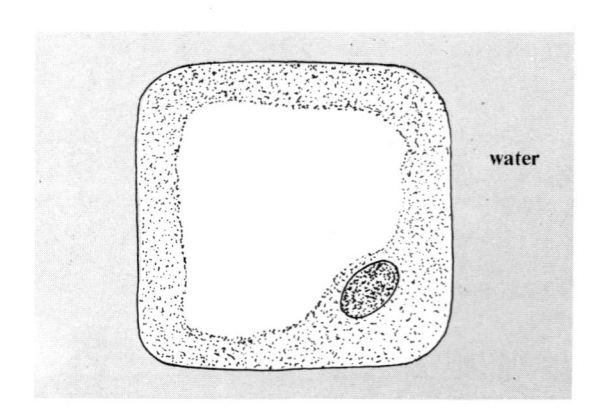

$$W \text{ is equal to } P_1 \text{ and } P_2 = 0$$
$$S = (P_1 - 0) - W$$
$$S = 0$$

Considering an isolated living vacuolated plant cell placed in concentrated salt solution

The salt solution should have a higher osmotic pressure than the cell sap. In this case the diffusion pressure of water in the external solution is less than that in the cell sap and water will pass from the vacuole through the protoplasm and cell wall to the external solution. As water leaves so the vacuole decreases in size and the contents start to shrink away from the wall of the cell, this condition is known as "incipient plasmolysis". The cell wall is completely permeable to the external solution and as the contents within the cell shrink away from the wall, so some of the external solution will pass through the wall to occupy the space left between the contents and the wall.

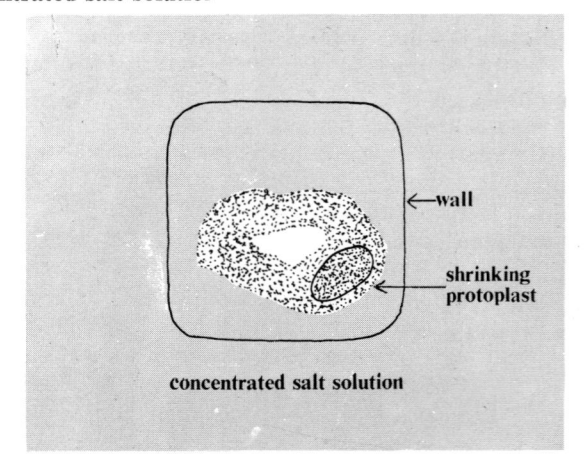

Thus for a cell placed in concentrated salt solution:—

$W = 0$
P_2 is greater than P_1
$S = (P_1 - P_2) - W$
S will equal a negative quantity and water
  will leave the cell

If a cell which is slightly plasmolysed is placed in pure water it is capable of absorbing water and returning to its original condition, de-plasmolysis has occurred.

$W = 0$
P_1 is greater than P_2 which is zero
P_1 is maximum
$S = (P_1 - P_2) - W$
S is therefore also maximum since S equals P_1

If the cell is left in the salt solution the cell contents eventually form a tight mass completely isolated from the cell wall, in which condition the cell is said to be fully plasmolysed. This condition is not reversible and if a cell which is fully plasmolysed is placed in water it cannot be "de-plasmolysed."

Considering living plant cells in situ in the intact plant.

When considering a cell as a part of an intact tissue one must take into account the surrounding cells. An isolated cell is affected only by the pressure from its own wall but a cell in a piece of tissue is affected by the pressure of the walls of the surrounding cells. This results in the development of **tissue tensions** which are the main source of rigidity in herbaceous plants. For example, if a fresh dandelion peduncle is split lengthways into about four pieces the strips immediately begin to curl outwards; the reason for this being that the epidermal cells are inextensible and therefore contained within a certain limited length, but the inner cells are capable of limited changes in size due to changes in turgidity and when they are fully turgid, as in a fresh stem, they are held under compression by the inextensible outer epidermis, but when this restraining layer is cut they are free to extend.

The movement of water from cell to cell in an intact tissue is not always as one would expect when considering the osmotic pressures of the cells, since the suction pressure of the cell depends not only on the osmotic pressure but also on the wall pressure, and is further complicated by intercellular pressures. One can determine roughly the direction of movement by comparing the gross suction pressures of the two cells.

Consider a cell A which has an osmotic pressure of six atmospheres and a wall pressure of two atmospheres, while cell B has an osmotic pressure of ten atmospheres and a wall pressure of eight atmospheres. Excluding the intercellular pressures it would appear that water would pass from B to A since the gross suction pressure of cell A is $6 - 2 = 4$, while the gross suction pressure of cell B is $10 - 8 = 2$ and water would move against the osmotic gradient.

Special effects of various solutes on plant cells

As mentioned on page 3 the cytoplasm is permeable to a limited number of solutes; some of these are capable of killing the cell or affecting the osmotic pressure of the cell sap once they have entered.

When a portion of tissue is placed in a solution and left for a few minutes, then removed and examined, it may be found that the cells are in the same condition as they were before they were immersed in the solution.

Alternatively it is possible that certain changes may have taken place depending on the solution into which the cells were placed.

1. If there is **no change** then it will be for one of the following reasons.
    a. the solution may be isotonic with the cell sap.
    b. the cytoplasm may be impermeable to the solvent.
    c. the solution may be a pure substance which cannot penetrate the cell, for example liquid paraffin.
    d. the solute (or pure substance) may be one that enters the cell very rapidly but is toxic thus killing the cell instantly therefore no change is visible, for example alcohol.

2. If there is **plasmolysis** followed by **recovery**

   The initial plasmolysis is caused because the solution is hypertonic to the cell sap. The recovery occurs because the cytoplasm is permeable to the solute which therefore passes through it by diffusion into the cell sap. The cell sap therefore increases in concentration until it reaches the point at which it is fractionally greater than the concentration of the external solution and de-plasmolysis of the cell contents commences. Thus the solution is hypertonic to the cell sap but the cytoplasm is permeable to the solute for example glycerol.

3. If there is **plasmolysis** but **no recovery**

   If plasmolysis of the cell contents occurs but there is no recovery then the solution must be hypertonic to the cell sap, thus causing the plasmolysis. The solute is either one which cannot penetrate the cytoplasm quickly enough to prevent the drastic plasmolysis which kills the cell, as for example with strong salt solution; or one which cannot penetrate the cytoplasm at all.

PASSAGE OF WATER INTO AND THROUGH THE PLANT

Entry of water into the plant

Plants generally obtain all the water which they require from the soil in which they are rooted, the water entering the plant through the root hairs which are confined to a specific region of the root. Except when the soil is waterlogged the soil water is held in the form of films surrounding the individual soil particles. It is also held in smaller spaces in the soil by capillary forces but it drains away from the larger spaces. There is little movement of the water held in the soil by capillary forces and rather than the water rising to the roots the roots tend to grow towards the source of water; this is a tropic response referred to as hydrotropism (see also page 114).

Since plants normally require large quantities of water, it is essential that they should be able to obtain adequate supplies from the surrounding soil. Since water enters the plant through the roots, the root system must be very extensive and capable of quick and efficient absorption of large quantities of water.

As mentioned in the first paragraph only part of the root system is capable of absorbing water, the older parts of the root having a corky outer covering which is almost completely impermeable to water. Maximum absorption occurs in a region a few millimeters in length just behind the root cap, referred to as the piliferous region; limited absorption can also take place in the meristematic region. It is in the piliferous region that the root hairs are located. These are small elongated epidermal appendages (see Book 1 page 16), and their function is to increase the surface area of the root in this region and to bring as much of this surface as possible into close contact with the soil particles, for quicker and more efficient absorption of water.

It is believed that there are two ways in which water enters the plant:—
1. by active absorption
2. by passive absorption

ACTIVE ABSORPTION

This refers to the entry of water into the plant due to the existence of an osmotic system between the soil solution which consists of a solution of various salts in water, and the cell sap within the vacuole in the root hair cell, the protoplasm acting as a semi-permeable membrane. The cell sap is of higher concentration than the soil solution and water enters the root hair by osmosis as described on page 3, there being a diffusion pressure deficit of water in the root hair cell with regard to the soil solution.

Water passes by osmosis from the root hair cell across the cortex, endodermis and pericycle of the root to the xylem of the central conducting strand. The cells towards the centre having a diffusion pressure deficit of water with regard to those near the outside. The water is passed from cell to cell via the protoplasts as explained previously. The passage of water from cell to cell in an intact tissue is also affected by tissue tensions but the general effect is the development of a gradient of decreasing diffusion pressure of water from the outside towards the inside.

Before the water can enter the central conducting tissue it has to pass through the endodermis. The **endodermis** consists of a single compact layer of specialised cells, the structure of which is dealt with in detail in Book 1 page 37. The principal feature of these cells is a band of specialised material in their radial and transverse walls which differs chemically from the rest of the cell wall. This band, which is impervious to water, is referred to as the Casparian strip. It prevents the passage of water through the endodermis via the cell walls. Thus the endodermis provides a barrier through which water can only pass via the cell contents. Due to the diffusion pressure deficit of water in the cells within the endodermis with regard to the cells outside it the direction of passage of water, by osmosis, through this layer is always towards the centre of the root. It also provides a barrier preventing the leakage of the water under pressure in the xylem back into the cortex.

This actual absorption of water can result in the development of a positive pressure in the xylem, referred to as **root pressure,** which can be measured experimentally (see page 30). It has been shown that root pressure is a variable and that some plants do not develop a measurable root pressure at all.

PASSIVE ABSORPTION OF WATER

This refers to the passive absorption of water through the walls of the root hair cells, caused by the movement of water through the plant in the transpiration stream. It is believed that as water is drawn up through the plant in the xylem vessels, so more water is drawn through the walls of the vessels from the surrounding cells. This in turn is replaced by water drawn from adjacent cells, again via the cell walls. In this way water is drawn across the root and since the root hair wall is in contact with the soil solution water is drawn from the soil solution through the root hair wall into the plant. Thus there is a 'column' of water extending from the leaves down through the xylem vessels to the root, then laterally across the root via the cell walls to the root hair wall. As water is removed at the top of the column in the leaves so more water is absorbed at the base to replace it, that is from the soil solution via the root hair walls.

It should be noted that water does not pass across the endodermis via the cell walls but via the cell contents. It is possible that the cytoplasm of endodermal cells is more permeable than that of the cortical cells. It is unlikely that water passes from cell to cell across the cortex simply by leakage through the cytoplasm of adjacent cells rather than by osmosis, but it is possible that some water might leak through the endodermis via the cytoplasm instead of passing through by osmosis.

Concept of the movement of water within the apoplast and symplast
A concept developed in a review by Salisbury in 1963, shows a clear division, in relation to water movement, between two areas of the plant, the apoplast and the symplast.

APOPLAST

This is the area throughout which free movement of water is possible. It includes the cell walls, the intercellular spaces and the xylem system. The result is an area extending from the piliferous layer and inner cortical cells of the root through to the cells of the leaves, but excluding the endodermis of the root. The cells of the endodermis have impervious suberised strips of lignin in their radial and transverse walls preventing free movement of water through these walls. The walls of some of the endodermal cells do not possess these strips and therefore permit free movement of water. These are referred to as passage cells. Water can only pass through the endodermis via the cytoplasm or via the walls of the passage cells.

SYMPLAST

This is the cytoplasmic system of the plant body. It is continuous throughout the living cells of the plant, the cytoplasm of adjacent cells being connected by plasmodesmata (see Book 1 page 6). The movement of water through the symplast is controlled by the phenomena governing the movement of water through a cytoplasmic system.

Passage of water into the xylem
Water is actively drawn into the cavities of the xylem vessels by osmosis, the fluid within the xylem vessels having a diffusion pressure deficit of water with regard to the surrounding pericyclic cells. It is also believed that water passes passively into the xylem as described above. The active passage of water into the xylem by osmosis may be aided by the fact that dissolved salts are absorbed by the root from the soil solution and are passed across the root via the living cell contents until, in the centre of the root, they tend to leak into the cavities of the xylem vessels, with the result that there is an increase in the concentration of the fluid within the xylem vessel cavities and an increase in the diffusion pressure deficit of water in these elements.

The movement of water through the plant
Water is passed from the root to all parts of the plant via the xylem tissue. The structure of the xylem is dealt with in detail in Book 1, page 41. Considering now the xylem of Angiosperm trees, as mentioned in Book 1 page 44. The conducting tissue of the xylem is composed of vessels, the walls of which are thickened and strengthened before the dissolution of the protoplast, preventing the vessels from collapsing as water is drawn up through their cavities. The specialised conducting elements, vessel segments, are arranged one above the other to form continuous tube-like structures extending from the xylem vessels within the leaves down to the xylem vessels within the roots.

Since these tubes are filled with a dilute solution of various substances dissolved in water, there are continuous columns of fluid, usually referred to as sap or just water, extending from the leaves to the root, within the cavities of the xylem vessels, each of these columns of water being held together by the cohesive force of the water particles and the adhesive force between the water and the walls of the xylem vessels.

Various theories have been put forward concerning the upward movement of water through the plant. The possibility of it being the result of **root pressure** is one. This is a pressure developed in roots due to the absorption of water from the soil and its passage across the root and into the xylem. It accounts for a small rise of water in some plants some of the time but it is not sufficient to carry water to the top of a tree, and some plants do not show a measurable root pressure.

The possibility that it is the result of **capillary action** within the vessels is another theory but this could only account at the most, for a rise of a few feet in the thinnest of the vessels. Any theory has to account for the large forces involved in raising water to the top of the tallest trees, which in some cases can be as much as 450 feet, and overcoming the friction between the water and the walls of the vessels and the speed of ascent, which in some cases is as much as 150 feet per hour.

The theory accepted at present is the **Dixon and Joly, Transpiration/Cohesion Theory.** This theory rests on the cohesive properties of water, and on the fact that there are continuous columns of fluid within the xylem vessels extending from the xylem in the root to the xylem in the leaves. A column of water in an airtight tube has great tensile strength, it can withstand a pull of 5,000 pounds per square inch, the tensile strength of a column of xylem fluid would be less than this, probably about 3,000 pounds per square inch but the movement of water to the top of the tallest trees would not need to withstand a pressure as great as this.

Given that this column of water could withstand the strain of being lifted through the plant it is now necessary to decide what force causes this movement. It is here that transpiration comes into the theory. **Transpiration** is the loss of water in the form of vapour from the aerial parts of the plant, the main site for transpiration being the surface of the leaves. For details of transpiration see page 9. Water in the form of water vapour is continually being lost from the moist surfaces of the cells in contact with the air space system of the leaf. As water is lost from these cells so it is replaced by water drawn from the adjacent conducting elements. Since there is a continuous column of water extending from these cells down to those in the root, as water is removed from the top of the column more water is drawn up from below to replace that removed.

It is considered that water rises through the plant as the result of several factors, the most important of which is **leaf suction,** which in some cases may be assisted by root pressure and capillary rise of water in the xylem vessels.

MINERAL REQUIREMENTS OF PLANTS

The presence of certain minerals is essential for the normal healthy growth of all plants. These minerals are obtained from the soil in which the plants are rooted, the minerals entering through the root system. In the case of aquatic plants or algae they are obtained from the surrounding water.

A deficiency in any of the essential minerals will have a definite and characteristic effect on the plant.

Six elements are required in fairly large quantities, these are listed below, with the effects which inadequate supplies would have on plants.

    nitrogen . . . deficiency results in chlorosis and dwarfing of plants.
    potassium . . . deficiency results in yellow edges to the leaves and the premature death of the plant.
    calcium . . . deficiency results in stunting of root and stem.
    phosphorus . . . deficiency results in dwarfing and dull dark green leaves.
    magnesium . . . deficiency results in chlorosis, the older leaves turn yellow but the veins remain green.
    sulphur . . . deficiency results in chlorosis.

There are a further seven elements which are also essential but are only required in small quantities these are:—

    iron . . . deficiency results in chlorosis, young leaves turn yellow, their veins remain green.
    copper . . . deficiency results in dieback of shoots.
    manganese . . . deficiency results in chlorosis and grey specks on leaves.
    zinc . . . deficiency results in malformed leaves.
    boron . . . deficiency results in brown heart disease.
    molybdenum . . . deficiency may result in slight dwarfing.
    chlorine } these are believed to be essential in some cases but the effects of deficiency are negligible.
    fluorine }

Sodium is sometimes absorbed by plants but is not essential except in halophytes. Although silicon is absorbed by grasses most of these appear to be able to live healthily without it. Aluminium may be absorbed sometimes having a toxic effect.

Processes involved in the uptake of minerals

There appear to be several stages involved in mineral uptake by cells.

First the ions diffuse from the soil solution into the intercellular spaces of the root apex, then they may become adsorbed onto the cell walls and subsequently diffuse through the cell wall and outer cytoplasmic membrane (plasma membrane) into the cytoplasm. If the plant has a low mineral content then these ions may accumulate in the vacuole of the cell, but if this already has a high concentration of ions then they may be transferred directly to the cytoplasm of the next cell probably via the plasmodesmata (intercellular protoplasmic connections).

Experimental work has shown that some minerals are more readily absorbed than others and that it is possible for the concentration of an ion to be greater within the cell vacuole than it was in the soil solution. It appears that the cytoplasmic membrane has some system by means of which the ions are selectively absorbed and that this absorption can take place against a concentration gradient, in which case energy will be required to drive the ions into the stronger vacuolar solution. It is believed that this energy is obtained from respiration since there is a close relationship between respiration and salt uptake.

Ions are able to diffuse freely between the intercellular spaces in the root apex and the external soil solution. Movement of ions through the cell walls and through the plasma membrane into the cytoplasm is also relatively easy but the transfer of ions from the cytoplasm to the tonoplast and into the vacuole requires energy and is believed to involve the use of carriers. It is believed that this is where the link between salt absorption and respiration lies.

TRANSLOCATION

As mentioned previously water travels through the plant in the xylem elements passing up from the root to the leaves and other aerial parts of the plant. There is also free passage of water throughout the apoplast area of the plant (see page 7).

Sugar travels from its source, which is usually the photosynthetic cells of the leaves but which may be a cell which is converting starch to sugar, to a region where it is required for use in a metabolic process, such as respiration or to a region where it may be stored for future use. These cells which utilise or store sugar occur in the roots, in the cortex, xylem rays, lateral and terminal meristems, flowers and fruits, and the sugar passes to them via the sieve tubes of the phloem. Some sugar passes to the roots and reacts with minerals to form amino acids.

Minerals enter the plant via the roots, where they may accumulate in the root cells or be passed to the xylem elements. Once in the xylem the minerals pass upwards in the xylem stream to the leaves, flowers, and fruits, where they are used. Any minerals not used may be passed back to the root.

Phloem transport can take place especially if the transpiration stream is slow. Movement laterally from the xylem to the phloem also takes place. Minerals are removed from the leaves before they fall; also from the petals of flowers, in which case they are passed on to the developing fruit.

TRANSPIRATION

Transpiration refers to the loss of water in the form of water vapour from the aerial parts of the plant. The main site for transpiration is the surface of the leaves, but it can also occur from any plant surface which is permeable to water and which is in contact with the surrounding air.

There are three forms of transpiration known as **cuticular, lenticular** and **stomatal**.

1. **Cuticular and lenticular transpiration**

The outer surface of the plant is covered by a superficial layer which is either epidermis or cork. Cork itself is almost completely impermeable to water but it is penetrated by lenticels through which water may be lost, (see Book 1 page 36). The amount of water lost in this way is negligible when compared with the amount lost by the plant as a whole. This loss of water vapour by the plant through the lenticels in the cork is referred to as **lenticular transpiration.**

Cuticular transpiration refers to the loss of water vapour through the cuticle. It occurs at any plant surface where the cuticle is very thin or not fully waterproof.

The outer surface of the epidermal cells which form the superficial layer of the primary plant body, is covered by a layer of cuticle (see also Book 1 page 23). This cuticle is composed of cutin, which is a complex mixture of fatty acids and their oxidation and condensation products. It provides an impervious external layer, the degree of impermeability depending to a certain extent on its thickness. In some cases the impermeability of the cuticle is increased by the deposition of a layer of wax on the surface of the cuticle. The cuticle covering young leaves is very thin and the amount of water vapour lost through it is high when compared with the amount lost by some desert plants where the cuticle is very thick and often covered by a layer of wax. In young leaves, before the stomata are fully open, transpiration can only take place through the cuticle, cuticular transpiration being the only form possible.

2. **Stomatal transpiration**

This is the loss of water vapour from the plant via the stomatal pores. These stomatal pores are situated mainly in the epidermis of the leaves but they are also found in limited numbers in the epidermis of herbaceous stems and on flower petals (see Book 1 page 31).

The leaves of the plant are the chief site for photosynthesis. For this reason they are specialised to provide a large area of moist cell surface in contact with the internal air spaces of the plant to facilitate quick and efficient exchange of gases between the individual cells and the air spaces. The stomata provide an essential link between the air spaces within the plant and the external atmosphere. However this arrangement also provides a large area of moist cell surface from which the evaporation of water vapour cannot be prevented. This water vapour diffuses freely through the air spaces due to a concentration gradient. There is a high concentration of water vapour near the surfaces of the cells and a low concentration in the sub-stomatal air cavities. As a result of this, water vapour diffuses from the cell surfaces towards the sub-stomatal air cavities where the concentration of water vapour is lower and from which it is lost to the external atmosphere via the stomatal pores.

THE ADVANTAGES TO THE PLANT OF TRANSPIRATION

It has been said that transpiration provides a cooling mechanism for the plant and that since it causes water to be drawn up through the plant it assists in the translocation of water and mineral salts; but it is doubtful if it serves any other advantage and it is generally accepted that it is a side effect resulting from the development of an efficient photosynthetic structure.

FACTORS AFFECTING THE RATE OF TRANSPIRATION

There are various factors which affect the rate of transpiration of a plant. Some are factors which would affect the rate of evaporation from any water surface, others are internal factors within the plant itself.

1. **The humidity of the atmosphere**

The rate of evaporation is higher in a dry atmosphere than in a humid atmosphere in which the air may be fully saturated with water vapour.

2. **The movement of the atmosphere, that is the affect of wind currents**

The presence or absence of wind, and its strength, will affect the rate of transpiration since it results in the movement of the air surrounding the aerial parts of the plant. When, as on a still day, there is very little or no movement of the air, then the air surrounding the plant has a fairly high water content and the rate of transpiration is low. If the air is continually changing due to a breeze of dry air, fresh dry air is continually coming into contact with the plant and the rate of transpiration is increased.

3. **The temperature of the surroundings**

Plants, unlike animals, attain the same temperature as their surroundings, thus an increase in the temperature of the surroundings results in an increase in the internal temperature of the plant. Since an increase in temperature increases the rate of transformation of liquid water to water vapour there is an increase in the rate of evaporation of water vapour from the moist cell surfaces within the leaves of the plant and therefore an increase in the rate of transpiration.

There may be a further increase in transpiration for warm air can hold more water than cold air. An increase in the temperature of the surroundings can result in an increase in the rate of transpiration especially if the air had previously been saturated with water vapour.

4. Atmospheric pressure

The rate of evaporation is increased by a decrease in the pressure of the surrounding air. Thus the rate of transpiration increases with an increase in rarity. This is one of the reasons for the modification of some alpine plants to decrease their rate of transpiration.

5. The opening and closing movements of the stomatal pores

It would appear that the rate of loss of water vapour from the plant via the stomatal pores is affected by the size of the pores and that any factor which affected the size of the pore would indirectly affect the rate of transpiration. It is thought that the size does not affect the rate unless the stomatal pores are less than half open, and that their diffusing capacity is not fully utilised when they are fully open. It has also been found that the size of the stomatal pore effective in reducing the rate of transpiration varies according to the velocity of the air movements in the vicinity of the leaves of the plant.

6. Intensity of light in the vicinity of the plant

As mentioned in Book 1, page 31, the opening and closing movements of the stomata are governed by the intensity of the surrounding light.

The size of the stomatal pores depends on the turgidity of the guard cells which in turn depends on the osmotic pressure of the sap of these cells. In the absence of light or in the presence of light of very low intensity the guard cells cannot photosynthesise, thus carbon dioxide which is produced as a by-product of respiration is not used and starts to accumulate causing the pH of the cells to drop to about 5·2. At this pH the enzyme diastase favours the conversion of sugar to starch. Thus osmotically active sugars in the cell sap are removed from solution and deposited as starch grains in the chloroplasts. The guard cells now have a lower osmotic pressure than the surrounding epidermal cells and so tend to lose water to them by osmosis. As the guard cells lose water so they start to collapse and close the stomata. As the intensity of the light surrounding the plant increases (for example at dawn), so the guard cells can commence photosynthesis and the carbon dioxide which had accumulated is used up, the pH of the cells rises and diastase now favours the conversion of starch to sugar. Since the sugar is osmotically active the osmotic pressure of the cell sap increases above that of the surrounding epidermal cells and water now passes into the guard cells from the surrounding epidermal cells by osmosis. The guard cells become turgid and the stomata open. Since the size of the stomatal pore can in some cases affect the rate of transpiration, (see paragraph 5 above) so the intensity of the light surrounding the plant can indirectly affect the rate of transpiration.

7. The water content of the leaf tissues

A decrease in the water content, and therefore also in the turgor of the guard cells results in the closing of the stomatal pores, and a general wilting of the leaves usually results in the stomata closing completely.

DISCUSSION OF STOMATAL TRANSPIRATION

The reason for the loss of water from the leaves of the plant is obvious when considering their structure. Since the leaf is the chief site for photosynthesis it is specialised to carry out this essential function efficiently.

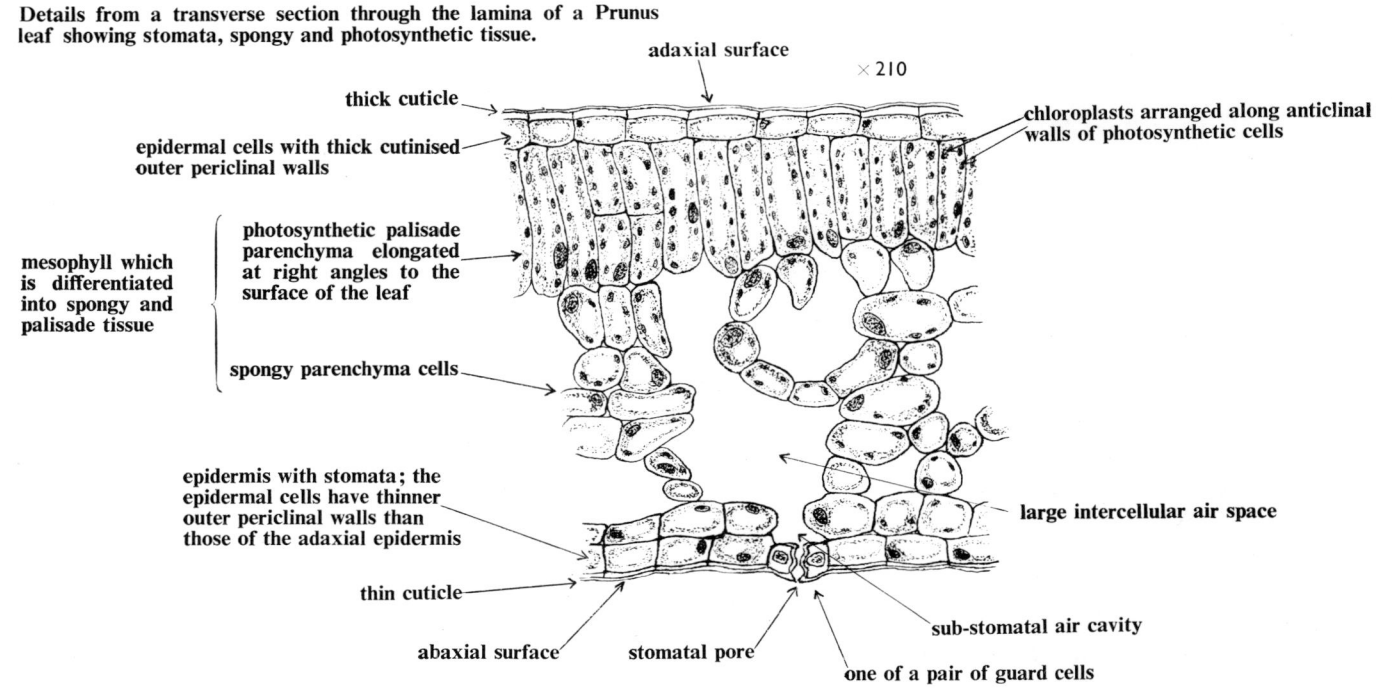

Details from a transverse section through the lamina of a Prunus leaf showing stomata, spongy and photosynthetic tissue.

×210

- adaxial surface
- thick cuticle
- epidermal cells with thick cutinised outer periclinal walls
- chloroplasts arranged along anticlinal walls of photosynthetic cells
- mesophyll which is differentiated into spongy and palisade tissue
- photosynthetic palisade parenchyma elongated at right angles to the surface of the leaf
- spongy parenchyma cells
- epidermis with stomata; the epidermal cells have thinner outer periclinal walls than those of the adaxial epidermis
- large intercellular air space
- thin cuticle
- sub-stomatal air cavity
- abaxial surface
- stomatal pore
- one of a pair of guard cells

The photosynthetic cells are in contact with the extensive air space system of the leaf, which is essential for the speedy exchange of gases required for, and produced during photosynthesis, since lack of carbon dioxide stops photosynthesis and the accumulation of oxygen in the cells will inhibit it. For the efficient exchange of these gases it is necessary that a large area of moist cell surface is exposed to the internal air spaces of the leaf. Unfortunately this also provides a large area from which the loss of water cannot be prevented without hindering the exchange of gases. The air space system is in contact with the external atmosphere via the stomatal pores for quick and rapid exchange of gases between the plant and its surroundings. From such a structure the loss of water vapour cannot be prevented.

CONSIDERING THE MOVEMENTS OF WATER VAPOUR FROM THE CELLS INTO THE AIR SPACES

The surfaces of the cell walls in contact with the air spaces are covered by a very thin layer of cuticle referred to as the internal cuticle. This layer has hydrophobic (water hating) properties. The cell walls in contact with the air spaces are penetrated by minute pores filled with water. This water tends to move outwards along the pores by capillary action, more water being drawn from the cytoplasm to replace it. This, in turn can be replaced by water drawn from the cell sap in the vacuole, or by water drawn directly from the xylem of adjacent conducting tissue. Water reaching the end of the pore is prevented from running out over the external surface of the cell by the hydrophobic layer of internal cuticle. Evaporation of water occurs from the meniscus at the end of each capillary column of water. As water evaporates at the surface of the meniscus so it is continually being replaced by the capillary action of the column of water in the pore. This evaporation of water at the surface of the cell results in an increase in the concentration of water vapour in the air spaces in this region, the concentration being much greater than that in the sub-stomatal air cavities. Thus water vapour diffuses away from the cell surfaces towards the sub-stomatal cavities along a gradient of decreasing concentration.

Water vapour in the sub-stomatal air cavities diffuses outwards through the stomatal pores if, as is usually the case, the concentration of water vapour in the external atmosphere is less than that in the sub-stomatal air cavity. If the air is very humid then the rate of diffusion of water vapour outwards, that is the rate of transpiration, is reduced but if the air is dry then the rate is increased.

The cells of the leaf from which the water vapour was originally lost to the air spaces replace the water by drawing more from the xylem of the nearest strand of conducting tissue with which they are in close contact. There is an extensive network of strands of conducting tissue extending throughout the leaves in all directions and all cells of the leaf are in close contact with part of the network and can easily draw water from the xylem of these strands.

REDUCTION OF THE RATE OF TRANSPIRATION IN PLANTS

There are various ways in which the rate of transpiration of a plant can be reduced. These modifications are seen in plants living in habitats in which it is necessary for them to conserve water, where water is either in short supply or in a form in which it cannot readily be absorbed or utilised by the plant (see also Book 1, page 88).

1. The development of a thick cuticle.
2. The development of thickening and cutinisation of the outer periclinal wall of the epidermal cells.
3. The development of a layer of wax over the cuticle.

These all reduce the amount of water vapour lost through the cuticle and therefore reduce cuticular transpiration.

4. The development of external air cavities and thick cuticular ridges around the stomatal pores.
5. The sinking of the stomata below the level of the general leaf surface.
6. The development of thick cuticular ridges on the guard cells around the stomatal pores.
7. The development of epidermal hairs, (see Book 1, page 27).
8. The rolling of the leaves so that the stomatal surface is partly enclosed.
9. The reduction in the size of the stomata.

These reduce the movement of air in or near the stomatal pores, cutting down the loss of water vapour by diffusion through the stomata, reducing stomatal transpiration.

OSMOSIS EXPERIMENTS

Experiment 1
Simple experiment to demonstrate osmosis

APPARATUS REQUIRED

semi-permeable membrane, this can be one of a number of things, for example:— parchment paper, a pig's bladder which has been washed and scraped, sausage paper or the shell membranes from a boiled egg or a piece of cellophane
thistle funnel
concentrated sugar solution
distilled water
beaker containing distilled water
white petroleum jelly
basin
thread
clamp and stand

METHOD OF PROCEDURE

Place the sugar solution in the basin then immerse the bulb of the thistle funnel in the sugar solution so that it enters and fills the bulb and rises a short distance up the stem of the funnel. While the bulb of the funnel is still immersed in the sugar solution take the membrane, place it across the open end of the bulb of the funnel and using the thread tie it securely in this position taking care that no air bubbles are trapped in either the bulb or the stem of the funnel, and that the joint between the bulb and membrane is water tight. Remove the funnel from the sugar solution and gently but thoroughly rinse the outside of the funnel and membrane in distilled water to remove any traces of the sugar solution. At this point the joint between the membrane and funnel can be covered with a layer of white petroleum jelly to ensure that it is completely watertight.

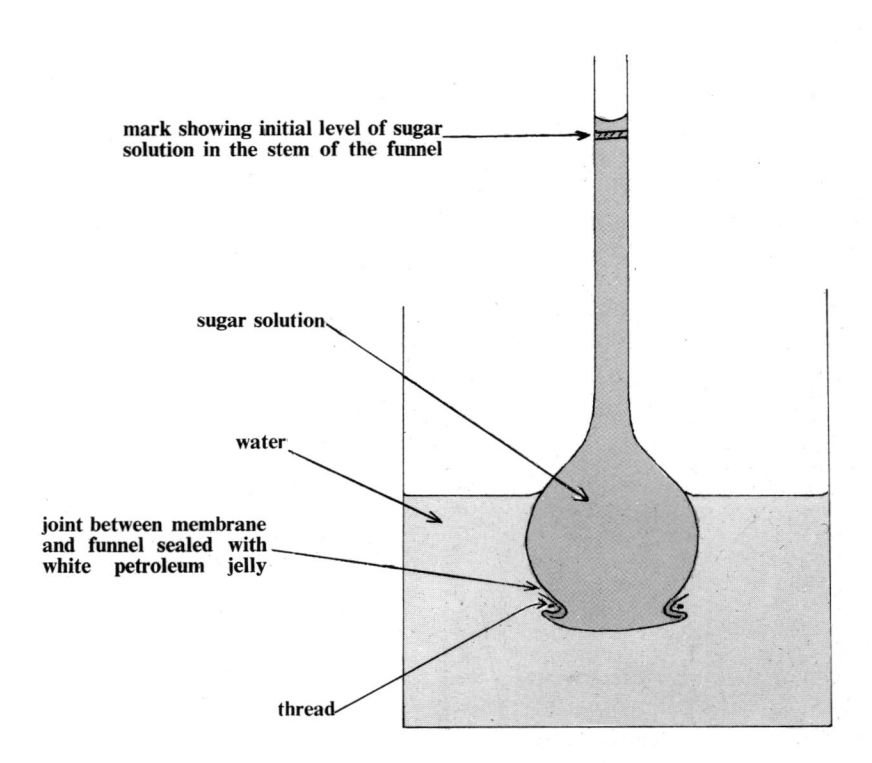

Immerse the bulb of the inverted funnel in a beaker of distilled water, supporting the stem of the funnel by a clamp and stand. Mark the initial level of the sugar solution in the stem of the funnel then leave the apparatus for several hours. Note any change in the level of the liquid in the stem of the funnel and determine whether or not osmosis has taken place.

Experiment 2
Microscopical demonstration of plasmolysis

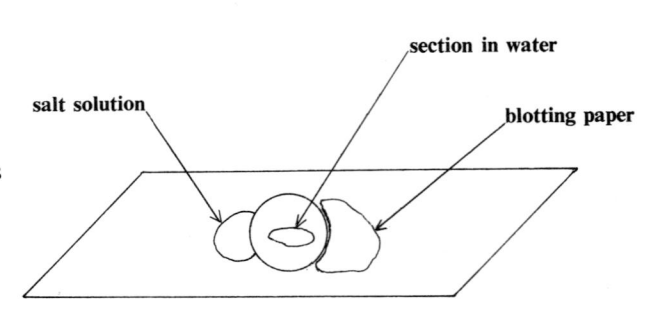

APPARATUS REQUIRED
fresh beetroot
salt solution; concentration equivalent to 30 atmospheres osmotic pressure (39·15 g of sodium chloride per litre)
distilled water
glass microscope slides and coverslips
small pieces of blotting or filter paper
microscope

METHOD OF PROCEDURE

Cut a thin section from the fresh beetroot and mount it in distilled water on a glass slide. Place the slide on the stage of the microscope, then make a high power drawing of a small group of cells from the beetroot tissue showing in particular the arrangement of their contents, noting especially their vacuoles which contain the red pigment anthocyanin in the cell sap. Place a few drops of 30 atmospheres salt solution on the slide to one side of the cover slip. Place a piece of blotting or filter paper against the opposite side of the cover slip, as shown in the diagram, drawing the water from beneath the coverslip. At the same time some of the salt solution will be automatically drawn under the coverslip, to take the place of the water which was removed. Remove the blotting paper to prevent the continued removal of the liquid under the coverslip. Observe the beetroot tissue as the salt solution is drawn in to surround it and the cell contents begin to plasmolyse. Note and make drawings of the various stages of plasmolysis. Then change the liquid surrounding the beetroot tissue by the same method as before, this time replacing the salt solution with water. Note any de-plasmolysis of the contents of the cells.

Note: This experiment can be repeated using any tissue whose cells have vividly coloured contents, for example surface sections of Fuschia, Coleus or Rhoeo discolor leaves or a complete Elodea leaf.
A tissue is considered to be plasmolysed when 50% of the cells are in a state of incipient plasmolysis.

Experiment 3
To determine the diffusion pressure deficit (suction pressure) of cells from the epidermis of an onion bulb scale

APPARATUS REQUIRED
fresh onion
6 watch glasses
scalpel
microscope and slides
salt solutions: a series of solutions of sodium chloride to give osmotic pressures of 5 to 30 atmospheres

g/litre sodium chloride	osmotic pressure of solution
6·525	5 atmospheres
13·050	10 atmospheres
19·575	15 atmospheres
26·100	20 atmospheres
32·625	25 atmospheres
39·150	30 atmospheres

METHOD OF PROCEDURE

Cut the onion into quarters then separate the scales and remove the epidermis from the inner concave surfaces. Place the salt solutions in the watch glasses, one solution to each watch glass. Take small pieces of the epidermal tissue and place one in each of the salt solutions.

After 15 minutes examine the pieces of tissue under the microscope for any sign of plasmolysis. Either place the watch glasses on the stage or mount the pieces of tissue on a microscope slide in some of the same salt solution in which they had been immersed. From your results determine the approximate diffusion pressure deficit (suction pressure) of the cells which you used.

Note: If no plasmolysis occurs in the solutions of 10, and 15 atmospheres and slight plasmolysis occurs in the solution of 20 atmospheres then the diffusion pressure deficit of the cells was between 15 and 20 atmospheres.
Thin sections of fresh beetroot or an Elodea leaf can be used instead of the onion epidermis if preferred.

Experiment 4

To determine the diffusion pressure deficit (suction pressure) of the cells of a potato tuber

APPARATUS REQUIRED
fresh potato
2 cork borers, 1 fairly wide and 1 narrow
salt solutions as in experiment 3 on page 14
6 shallow glass or plastic dishes
6 test tubes
test tube stand
2 scalpels or razors

METHOD OF PROCEDURE 1, USING DISCS OF POTATO TISSUE
Remove the peel from the potato, then using the wide cork borer cut several cylindrical pieces from the potato tissue. Cut each cylinder into several discs about 2 mm thick. Place samples of the salt solutions, one in each of the 6 glass dishes and place about 3 discs in each of the salt solutions, having first measured and noted the diameter of each of the discs and whether they are firm or flabby. Leave the discs in the salt solutions for about 30 minutes then remove them and re-measure their diameters noting any change in rigidity. Record your results in tabloid form as shown below, the diameter in each case being the average for the 3 in that solution. From your results determine the approximate diffusion pressure deficit (suction pressure) of the cells from the potato tuber.

Strength of salt solution	original diameter of discs	final diameter of discs	change in diameter of discs + or —
5 atmospheres			
10 atmospheres			
15 atmospheres			
20 atmospheres			
25 atmospheres			
30 atmospheres			

METHOD OF PROCEDURE 2, USING CYLINDERS OF POTATO TISSUE
Using the narrow cork borer cut several cylinders of potato tissue from a fresh peeled potato. Place the cylinders of potato tissue side by side and cut them all to the same length using 2 parallel blades. Place samples of the salt solutions, one in each of the 6 test tubes, then after measuring and noting the length of each of the cylinders of potato tissue place one in each of the test tubes making sure that each is fully immersed in the salt solution. After about 30 minutes remove the cylinders of potato tissue, re-measure their lengths and note any change in their rigidity. Record your results in tabloid form as shown below, and from these determine the approximate diffusion pressure deficit (suction pressure) of the potato cells.

Strength of salt solution	original length of cylinders	final length of cylinders	change in length of cylinders + or —
5 atmospheres			
10 atmospheres			
15 atmospheres			
20 atmospheres			
25 atmospheres			
30 atmospheres			

Experiment 5
To determine the diffusion pressure deficit (suction pressure) of cells from a potato tuber by change in weight of discs of potato tissue

APPARATUS REQUIRED
fresh potato
wide diameter cork borer and scalpel
salt solutions as in experiment 3 on page 14
6 shallow glass or plastic dishes
blotting paper or filter paper

METHOD OF PROCEDURE
Place the salt solutions in the dishes, one solution in each dish. Use the cork borer to cut several cylinders of potato tissue from the peeled potato, then cut the cylinders into discs about 2 mm thick. Dry the discs on blotting paper or filter paper then weigh them in sets of 6 and place one set in each salt solution. Leave them for about 1 hour then dry and re-weigh the discs noting any change in weight. Record your results, in tabloid form as shown in the previous experiment then determine the approximate diffusion pressure deficit (suction pressure) of the potato cells.

Experiment 6
To determine the diffusion pressure deficit (suction pressure) of potato tissue by measurement of changes in the specific gravity of the external solution

APPARATUS REQUIRED
fresh potato
narrow cork borer
knife or scalpel
16 test tubes (8 wide, 8 narrow)
2 test tube stands
8 dropping pipettes
neutral red solution
sucrose solutions of the following molar concentrations:— 0·1, 0·2, 0·3, 0·4, 0·5, 0·6, 0·8, 1·0

METHOD OF PROCEDURE
Place 10 ml portions of the sucrose solutions into the set of wide test tubes. Label each tube according to the concentration of the solution within it.

Using a narrow cork borer, cut cylinders of peeled potato tissue then slice these into discs which are approximately 2 mm thick. Place the same number of discs (15) in each tube and leave for 1-2 hours shaking the tubes at intervals.

Take 5 ml portions of the sucrose solutions, place them in the set of narrow test tubes and label each tube. Add a few grains of neutral red to each tube to colour the solutions.

After 1-2 hours remove the discs, retaining the solutions. Using the dropping pipettes remove some of the coloured solution from each of the second set of tubes. Wipe the outside of each pipette then immerse the end of it in the corresponding solution in the set of wide test tubes, which previously contained the potato discs. Very gently squeeze the bulb so that a drop of coloured solution is squirted into the middle of the solution in the test tube. Gently remove the pipette and observe the direction and speed of movement of the coloured drop.

Determine the diffusion pressure deficit (suction pressure) of the cells of the potato, using the table below.

RELATIONSHIP BETWEEN MOLAR CONCENTRATION AND OSMOTIC PRESSURE

molar concentration of sucrose	osmotic pressure of sucrose solution at 20°C
0·1	2·6 atmospheres
0·2	5·1 atmospheres
0·3	7·7 atmospheres
0·4	10·2 atmospheres
0·5	12·9 atmospheres
0·6	15·5 atmospheres
0·8	21·1 atmospheres
1·0	26·9 atmospheres

see note on following page

Note: Solutions which were **hypertonic** (osmotic pressure stronger than the suction pressure of the potato cells) will have taken up water from the discs of potato. Since such solutions will have become diluted their specific gravities will have decreased and they will have become less dense. In these solutions the coloured drops will sink. **Hypotonic** solutions (osmotic pressure weaker than the suction pressure of the potato cells), will have lost water to the potato discs and their specific gravities will have increased, the solutions becoming more dense. In these solutions the coloured drops will rise. The solution which is **isotonic** (the same strength) with the suction pressure of the potato cells will show no change in specific gravity, since it will not have gained or lost water. In this case the drop will remain stationary.

If the potato discs are weighed before and after immersion in the sucrose solutions, then this experiment may be combined with the determination of the diffusion pressure deficit by change in weight of the discs, as in experiment 5 on page 16.

Experiment 7
To determine the effect of various treatments on the permeability of beetroot cells

APPARATUS REQUIRED
fresh beetroot
scalpel and wide diameter cork borer
8 test tubes
test solutions:— distilled water
          0·3N sodium chloride solution
          0·3N calcium chloride solution
          half saturated chloroform water
          0·1N acetic acid
          0·1N sodium hydroxide solution

METHOD OF PROCEDURE
Cut 8 discs (or cubes) of fresh beetroot tissue, wash them thoroughly and treat as follows:—

    disc 1: place in distilled water in a test tube and leave
    disc 2: place in distilled water in a test tube and boil for 1 minute
    disc 3: place in a test tube with some 0·3N sodium chloride solution
    disc 4: place in a test tube with some 0·3N calcium chloride solution
    disc 5: place in a test tube containing a mixture of 9 parts sodium chloride and 1 part calcium chloride
    disc 6: place in a test tube with some half saturated chloroform water
    disc 7: place in a test tube with some 0·1N acetic acid
    disc 8: place in a test tube with some sodium hydroxide solution

Observe the relative rates at which the pigment leaves the beetroot tissue in each case, by noting the time taken for the density of the pigment in the solutions in the test tubes to reach a standard. This standard is made by boiling some discs in water then diluting the coloured solution obtained. If this standard solution is too concentrated then the time taken in some cases to reach the standard would be rather long.

Note: A full explanation should be included with your results.

Experiment 8
To determine the effect of various solutes on cells from the epidermis of an onion bulb scale

APPARATUS REQUIRED
epidermis from the concave side of onion bulb scales
microscope slides
microscope
test solutions:— molar potassium nitrate solution
          half saturated chloroform water
          molar potassium thiocyanate
          ethylene glycol
          50% ethyl alcohol
          absolute alcohol
          dilute glycerol
          concentrated glucose solution
          urea
          distilled water

METHOD OF PROCEDURE

Mount pieces of the epidermis from the concave side of onion bulb scales on glass microscope slides in the following solutions:—

1. molar potassium nitrate solution
2. molar potassium nitrate after placing in half saturated chloroform water for 1 minute
3. molar potassium nitrate solution after placing in half saturated chloroform water for 5 minutes
4. molar solution of potassium thiocyanate
5. ethylene glycol
6. ethyl alcohol 50%
7. absolute alcohol
8. dilute glycerol
9. concentrated glucose solution
10. urea
11. distilled water

Observe carefully any changes which may take place in the cell contents and determine why these changes have taken place. If there is plasmolysis followed by recovery note the time taken.

Note: A full explanation should be included with your results.

Experiment 9

To demonstrate the passage of water through plant tissue

APPARATUS REQUIRED
potato
glass dish
distilled water
knife or scalpel
lump of sugar

METHOD OF PROCEDURE

Remove the peel from both ends of the potato. Hollow out a fairly large cavity from one end and remove about $\frac{3}{4}$ inch of peel all round the other end. Fill the dish with distilled water and stand the potato in the dish making sure that the portion from which the peel has been removed is completely immersed in the water. Place the lump of sugar in the cavity at the other end then leave the experiment for several hours. Note any change in the sugar and draw your conclusions as to how and why this change has taken place.

Note: Solid sugar can act as an osmotic solution in the same way as does a strong sugar or salt solution.

Experiment 10

To determine the effect of pH and temperature on the permeability of cells from beetroot

APPARATUS REQUIRED
fresh beetroot
scalpel
cork borer
3 beakers
3 bunsen burners, tripods and water baths
dilute sodium carbonate solution (M/1000)
dilute hydrochloric acid (M/1000)
distilled water

METHOD OF PROCEDURE

Place some distilled water in one beaker, dilute sodium carbonate solution in the second and dilute hydrochloric acid in the third. Cut 3 discs of beetroot tissue of equal size wash them, then place one in each of the beakers. Heat the beakers slowly, on the water baths, over the bunsen burners and note the temperature in each case at which the pigment is released from the beetroot cells.

From your results determine the effect of pH on the permeability of beetroot cells.

Experiment 11

To determine the minerals essential for healthy plant growth using water culture solutions

APPARATUS REQUIRED

culture solutions, see note at end of experiment
seedlings of barley, wheat or broad beans
large glass jars each fitted with a waxed cork bored with holes to take seedlings
black paper
cotton wool

METHOD OF PROCEDURE

Fill each of the glass jars with a different culture solution. Fit the seedlings into the holes in the corks using cotton wool to hold them in position. Insert the corks in the jars then surround each jar with a piece of black paper to exclude light. After labelling the jars according to their contents, place them in a well lit position and leave them for several weeks.

Inspect the seedlings at regular intervals and record their growth, and any other effects noticeable. Aerate the solutions daily to ensure adequate oxygen supply to the roots.

Note: There are various "formulae" for culture solutions—one of which is listed below, this is Knop's culture solution.

The complete culture solution consists of:— 1000 ml distilled water
0·8 g calcium nitrate
0·2 g magnesium sulphate
0·2 g acid potassium phosphate
0·2 g potassium nitrate
a few drops of ferric phosphate or ferric chloride

To determine which minerals are essential for healthy plant growth it is necessary to make a series of solutions omitting a different element in each case, as below.

1. To omit nitrogen use calcium sulphate and potassium sulphate instead of the nitrates.
2. To omit potassium use sodium salts instead of the potassium salts.
3. To omit calcium use sodium nitrate instead of calcium nitrate.
4. To omit magnesium use sodium sulphate instead of magnesium sulphate.
5. To omit iron leave out the ferric salt.
6. To omit sulphur use magnesium nitrate instead of magnesium sulphate.

Experiment 12

To demonstrate tissue tensions using a dandelion peduncle

APPARATUS REQUIRED

fresh dandelion peduncle
scalpel

METHOD OF PROCEDURE

Take the fresh dandelion peduncle and cut it into 4 lengthways, note any curvature which takes place in the cut pieces and the direction of curvature. Draw your conclusions as to why this has taken place and why it does not when the peduncle is cut across.

Note: If a hollow cylinder with a 2 layered wall, in which the inner layer is held in compression and prevented from expanding by the outer layer, is slit lengthways into several pieces, the with-holding power of the outer layer is removed and the inner layer is allowed to expand.

Experiment 13
Experiment to demonstrate tissue tensions using a piece of Elder shoot

APPARATUS REQUIRED
piece of Elder shoot of the current year's growth
scalpel
ruler

METHOD OF PROCEDURE
Take the piece of Elder shoot which should be about 2 or 3 inches long and accurately measure its length. Using the scalpel make 4 shallow longitudinal slits in the outer tissues, then quickly peel them off from around the pith leaving the pith free. Re-measure the lengths of the various pieces **at once,** and decide why any changes take place.

TRANSPIRATION EXPERIMENTS

Experiment 14
To demonstrate that water vapour is given off by a leafy shoot

APPARATUS REQUIRED
potted plant which should have been well watered previously
rubber, or thick polythene bag
2 bell jars
2 glass plates
white petroleum jelly
thin string

METHOD OF PROCEDURE
After making sure that the soil is damp place the pot, containing the plant, in the rubber bag, gather together the neck of the bag so that it completely covers the pot, soil surface and base of the plant, tying it firmly around the base of the shoot of the plant. Make sure that the joint between the neck of the bag and the base of the shoot is airtight by sealing it with white petroleum jelly. Place the covered pot with plant on one of the glass plates and cover it with a bell jar, sealing the joint between the glass plate and bell jar with white petroleum jelly. Place the other bell jar on the second glass plate sealing it in the same way as the first.

Place both sets of apparatus in a warm sunny place and leave for 2 or 3 hours, noting any changes which occur within the bell jars.

Note: If a plant within a confined space gives off water in the form of vapour to the surrounding air through transpiration, then the air in that space will quickly become saturated with water vapour, some of which will condense as liquid water on the inner surface of the container.

Experiment 15
Simple experiment to demonstrate from which of the two surfaces of a leaf most water vapour is lost

APPARATUS REQUIRED
fresh shoot with at least 4 leaves
white petroleum jelly
tripod
4 pieces of cotton

METHOD OF PROCEDURE
Remove 4 leaves from the shoot, attach a piece of cotton to each leaf stalk, then treat the four leaves as follows:—

  first leaf: cover the abaxial surface with white petroleum jelly
  second leaf: cover the adaxial surface with white petroleum jelly
  third leaf: cover both surfaces with white petroleum jelly
  fourth leaf: leave untouched

Suspend the leaves, by means of the pieces of cotton, from a tripod. Leave them for about 1 week, examining the leaves every day and noting any changes.
From your results determine from which surface of the leaf water is lost.

Note: If each of the leaves is weighed, on a hanging balance, at the beginning of the experiment after treatment with white petroleum jelly, then re-weighed after 24 hours, the loss in weight is equal to the weight of water transpired by the leaf.

Experiment 16
To compare the rate of loss of water in the form of vapour, from the two surfaces of a leaf, using cobalt chloride paper

APPARATUS REQUIRED
potted plant
cobalt chloride paper (blue colour)
glass slides
rubber bands or clips

METHOD OF PROCEDURE
Choose a sturdy healthy leaf, leaving it attached to the plant. Place a piece of cobalt chloride paper (blue) against the upper surface of the leaf and secure it in this position using 2 glass slides placed one on either side of the leaf and held in position with rubber bands or clips.

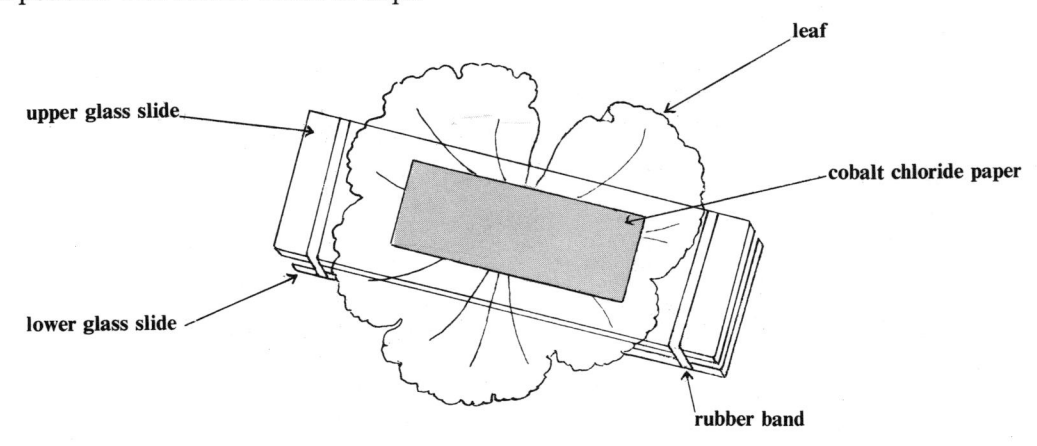

Select a second leaf, place a piece of cobalt chloride paper against the lower surface of the leaf holding it in position with glass slides as above.
Leave the experiment, inspecting from time to time. Note any change in the colour of the cobalt chloride paper, and the time taken for this change to occur. Compare the results obtained from the 2 positions of cobalt chloride paper.

Note: Cobalt chloride paper is used to detect the presence of moisture. When it is completely dry it is a deep blue colour, but when in the vicinity of a surface from which water vapour is evaporating, it gradually loses its blue colouration. The rate at which the colour is lost is related to the rate of evaporation from that surface.

Experiment 17

To compare stomatal and cuticular transpiration

APPARATUS REQUIRED
potted plant
white petroleum jelly or soft white wax
2 small bell jars
2 small tubes containing calcium chloride
2 U-tube manometers containing liquid paraffin or glycerine, one tube should have an additional backward bend
2 rubber bungs to fit the bell jars, one with a single hole through it, the other with two holes
2 clamps and stands
cotton
glass rod

METHOD OF PROCEDURE
Weigh accurately both tubes of calcium chloride. Tie a piece of cotton around one of the tubes. Pass the other end of the piece of cotton through one of the holes in the 2 holed bung, holding it in position with a glass rod passed through the same hole. Insert the bung in the neck of one of the bell jars so that the tube of calcium chloride hangs inside the bell jar. Connect the bell jar with the U-tube manometer which has an additional backward bend. Place the bell jar firmly but gently against the upper surface of a large smooth leaf of the plant. The leaf should be large enough to completely cover the opening of the bell jar. Seal the joint between the leaf and the jar with white petroleum jelly or soft white wax. Support the bell jar using a clamp and stand.

Invert the second bell jar, insert the single holed bung in the neck and insert one limb of the second U-tube manometer in the hole. Place the second tube of calcium chloride inside the jar then place the inverted bell jar against the lower surface of the leaf again sealing the joint with white petroleum jelly or soft wax. Support the bell jar using a clamp and stand.

Mark the levels on the manometers, leave for 15 minutes during which time the water vapour initially present in the bell jars is absorbed by the calcium chloride, causing a slight drop of pressure which is registered on the manometers. If no drop in pressure is registered then there must be a leak which must be sealed before the experiment can continue.

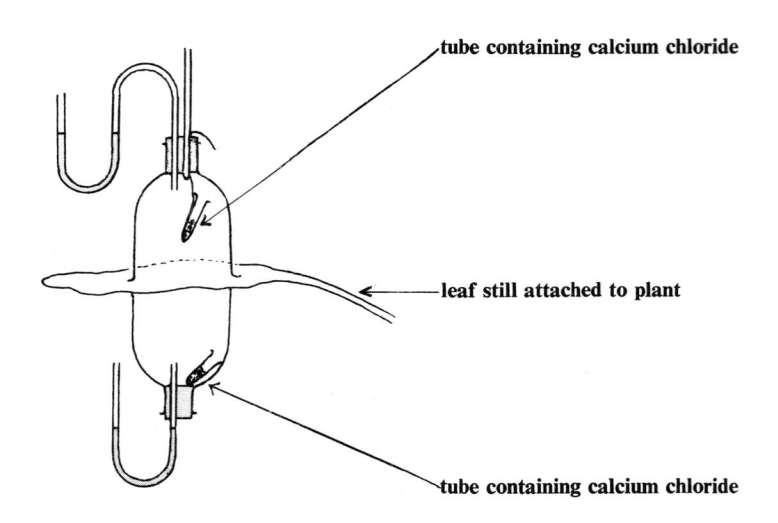

After making certain that the apparatus is airtight leave it for several hours, noting the length of time. Then re-weigh the tubes of calcium chloride. The increase in weight in each is equal to the amount of water lost by that portion of the leaf surface enclosed by the bell jar.

Calculate the amount of water lost through the surface in milligrams per square centimetre of leaf surface, per hour.

Note: Check, by microscopic examination, that the stomata are confined to the lower surface of the leaf. The upper surface can therefore be considered to give a measurement of cuticular transpiration while the lower surface gives a measurement of stomatal transpiration plus cuticular transpiration.

Experiment 18
To compare the stomatal densities of two leaves

Note: When the air spaces of a leaf are filled with liquid the leaf becomes dark and translucent. The penetration of a leaf by a liquid can therefore usually be followed visually.

APPARATUS REQUIRED
leaves to be investigated
xylol
dropping pipette

METHOD OF PROCEDURE
Select 2 leaves with greatly differing stomatal densities and apply a drop of xylol to each (xylol penetrates more easily than water).

Note whether there is any difference in the speed at which the liquid penetrates the 2 leaves, and decide why this should be. Determine from your results which leaf must have the greater density of stomata.

Experiment 19
Examination of stomata to determine their frequency per square centimetre of leaf surface

APPARATUS REQUIRED
leaf to be investigated
colourless nail varnish
microscope
fine forceps
graduated slide

METHOD OF PROCEDURE
A convenient way of examining the arrangement of the stomata is to prepare casts of the leaf surface using cellulose acetate (nail varnish is cellulose acetate plus colouring).

Paint a thin layer of nail varnish on the upper and lower surfaces of the leaf to be examined. Allow this to dry then strip off the film of varnish using the forceps. Examine this strip under the microscope with the side that was next to the leaf uppermost. Determine the frequency and arrangement of the stomata within a given area. Use a graduated slide to estimate the area of the microscope field of view. Calculate the frequency of stomata per square centimetre of leaf surface.

Note: This method is not suitable for leaves possessing a large number of epidermal hairs.
Since the frequency of stomata decreases down the shoot it is useful to compare leaves from the top and the bottom of the shoot.

Experiment 20
To demonstrate the use of a porometer

Note: A porometer is a piece of apparatus designed to measure the amount of gas diffusing from a leaf surface.

APPARATUS REQUIRED
porometer
potted plant
white petroleum jelly
beaker of coloured liquid, (paraffin)
cork borer
thin piece of gelatine: this is formed by pouring warm 30% gelatine into a petri dish to form a layer 2 to 3 mm
   thick, then leaving it until set

METHOD OF PROCEDURE
The porometer consists of a small cup shaped funnel with a flattened rim and a short stem which is connected by a piece of rubber tubing to a T-shaped piece of glass tubing. The funnel is connected to one end of the cross limb, a short piece of straight glass tubing is attached by rubber tubing to the other end. This piece of rubber tubing has a spring clip. The vertical limb of the T-piece is long and dips into the beaker of coloured liquid.

diagram on following page

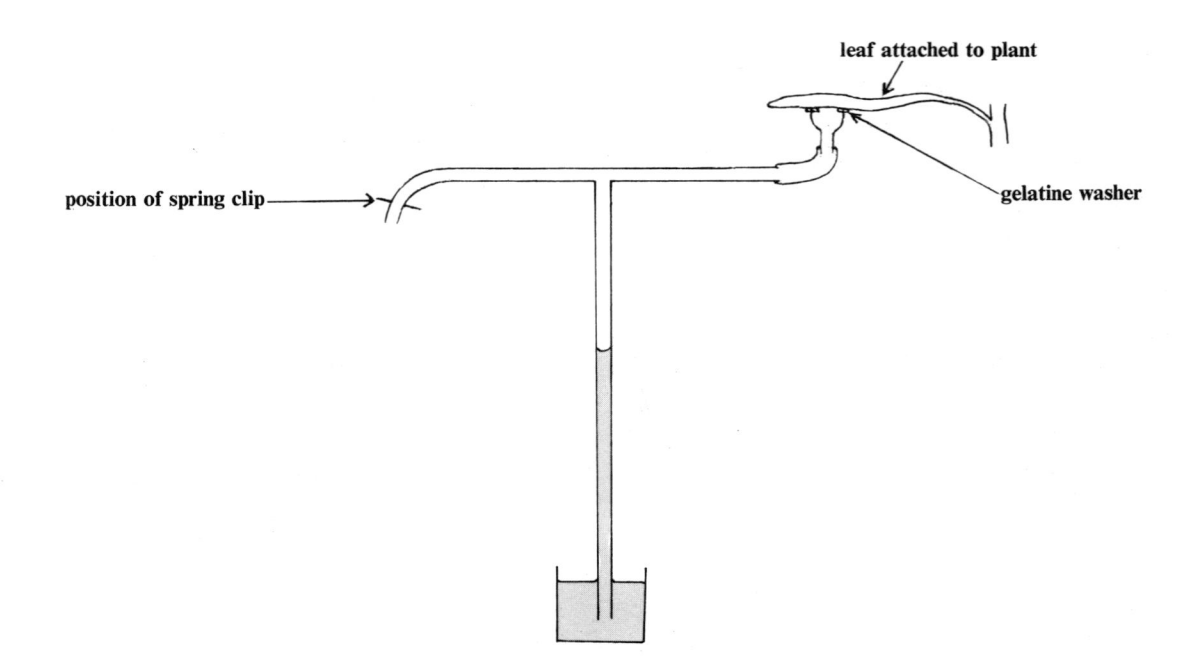

Select a leaf on the potted plant. Cut a piece of gelatine from the thin layer large enough to cover the funnel. Using a cork borer remove a piece about 4 mm diameter from the centre of the piece of gelatine which can then function as a washer. Push the open end of the funnel into the washer, then smear white petroleum jelly over the upper surface of the washer and push it firmly but gently onto the abaxial surface of the leaf. The white petroleum jelly acts as a seal between the leaf and the gelatine washer.

Open the clip and suck air from the apparatus; this raises the level of the liquid in the central limb. Close the clip and mark the level of the liquid. Leave the apparatus for several hours and note any change in the level of the liquid.

Note: This experiment may be repeated under different conditions of temperature and humidity but before proceeding with measurements under the new conditions allow at least 15 minutes for the plant to adjust to the new conditions.

In this way it is possible to compare the rate of diffusion of gases through the stomata under different conditions and therefore the extent of stomatal opening under different conditions. By using 2 porometers and attaching one to the abaxial and the other to the adaxial surface of the leaf it is possible to compare the distribution of stomata. Similarly comparisons may be made between numbers of stomata in different types of leaves.

This experiment also demonstrates the continuity of the air space system within the leaf since the only way in which air can enter the apparatus is through the leaf surface, if it continues to pass through then obviously it must be drawn through the leaf from other air spaces.

Experiment 21

To determine the volume of the air space system of a leaf

APPARATUS REQUIRED
leaves to be examined
'Buchner' flask, with tightly fitting rubber bung
filter pump
water

METHOD OF PROCEDURE
Take a few of the leaves to be examined and weigh them together accurately. Place them in the 'Buchner' flask and cover them with water. Insert the bung in the neck of the flask and connect the side limb to the filter pump, which should be connected to the tap. Turn the tap on and "vacuum infiltrate" the leaves with water. Remove them from the flask, blot them gently and re-weigh them quickly.

The increase in weight is due to the filling of the air spaces with water instead of air.

Since 1 g of water has a volume of 1 cm^3 at 4°C, then for this experiment the weight of water in grams can be taken to be equal to the volume of water in cubic centimetres (cm^3), and hence the volume of the air space system within the leaf.

Experiment 22

To demonstrate transpiration by the loss of weight of a leafy shoot

APPARATUS REQUIRED
freshly cut leafy shoot
wide cylindrical specimen tube or glass boiling tube
rubber bung, with 2 holes, to fit tube
piece of glass tubing to fit one of the holes in the bung
oil with dropping pipette.

METHOD OF PROCEDURE
Fill a deep sink or bowl with tap water. Fit the piece of glass tubing through one of the holes in the rubber bung then place the bung and specimen tube in the sink of water. Immerse the cut end of the leafy shoot in the water, taking care to prevent the leaves of the shoot from getting wet. Re-cut the stem of the shoot at a distance of at least one inch from the cut end. Keeping both the cut end of the shoot and the bung under water pass the cut end of the shoot through the second hole in the bung, then insert the bung in the neck of the tube (also immersed in water), making sure that the specimen tube and glass tubing are full of water and that there are no air bubbles trapped within them.

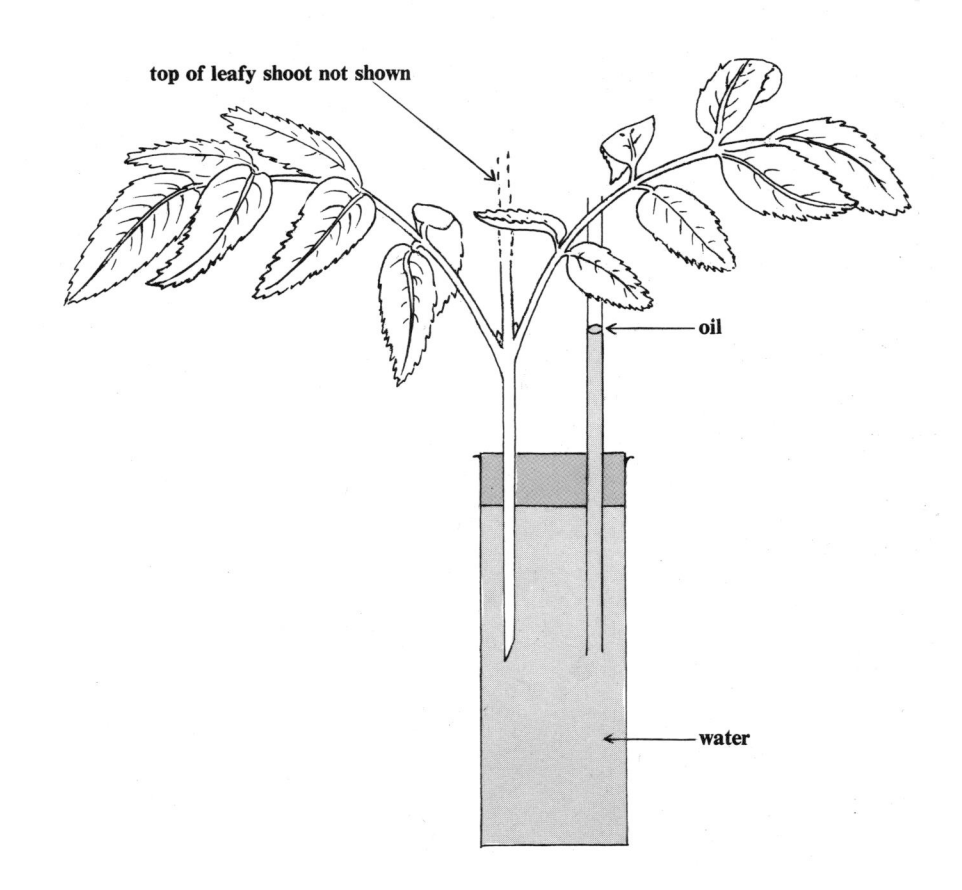

Remove the apparatus from the water and dry the outside of the specimen tube and glass tubing. Place a drop of oil on top of the water in the glass tube using a dropping pipette. Weigh the complete apparatus then place it in a warm well aerated position. Re-weigh after about 1 hour, noting the time exactly.

From your results, calculate the amount of water lost from the apparatus in the given time (1 hour) which, since water can only be lost through the leaves, is therefore equal to the weight of water transpired by the shoot. Calculate the total area of leaf surface as in experiment 23 on page 26 and determine the rate of transpiration in form of millilitres of water per square centimetre of leaf surface per hour, ml/cm²/h.

Note: There will be a slight experimental error due to the fact that the shoot will take time to become adjusted to its new surroundings before transpiration settles to a steady rate. This may be reduced by placing the apparatus in a warm well aerated position immediately after its removal from the sink of water and drying, then allowing it to become adjusted to these surroundings and then weighing it. Alternatively the apparatus may be re-weighed several times at regular intervals of 1 hour and the average weight taken for calculation of the rate of transpiration.

Experiment 23

To measure the rate of transpiration of a leafy shoot using a potometer

Note: A potometer is a piece of apparatus specially designed to measure the rate of transpiration of a leafy shoot by measuring the rate at which water is taken up by that shoot. It is based on the principle, that as water is lost from the shoot by transpiration so an equivalent quantity of water is taken up by the shoot to replace that which is lost. Therefore by calculating the transpiring surface area of the leaf and measuring the amount of water taken up by the shoot, it is possible to calculate the rate of transpiration per unit area of leaf surface.

There are various types of potometer all of which function in a similar manner but which vary in their accuracy. No potometer is 100% accurate, each apparatus having its own standard error which can be calculated, see experiment 27 on page 28.

APPARATUS REQUIRED

potometer as shown in the diagram below, fitted with a rubber bung with a single hole for shoot
small glass dish filled with water
freshly cut leafy shoot, the diameter of the stem being slightly greater than the diameter of the hole in the bung

METHOD OF PROCEDURE

Fill a deep sink or bowl with tap water. Remove the bung from the potometer then immerse bung and potometer in the sink. Taking care not to get the leaves wet, place the cut end of the leafy shoot in the water and re-cut the stem at least 1 inch away from the cut end. Fix the cut end of the shoot through the hole in the rubber bung and insert this in position in the potometer, checking that no bubbles of air are trapped in the apparatus. Remove the potometer from the sink and check that all the joints are airtight. If necessary cover the joints with a film of white petroleum jelly. Place the end of the capillary tube in the water in the small glass dish. Place the apparatus in a warm well aerated position and leave for about 15 minutes to adjust to the conditions of its surroundings, and to reach a steady rate of transpiration again.

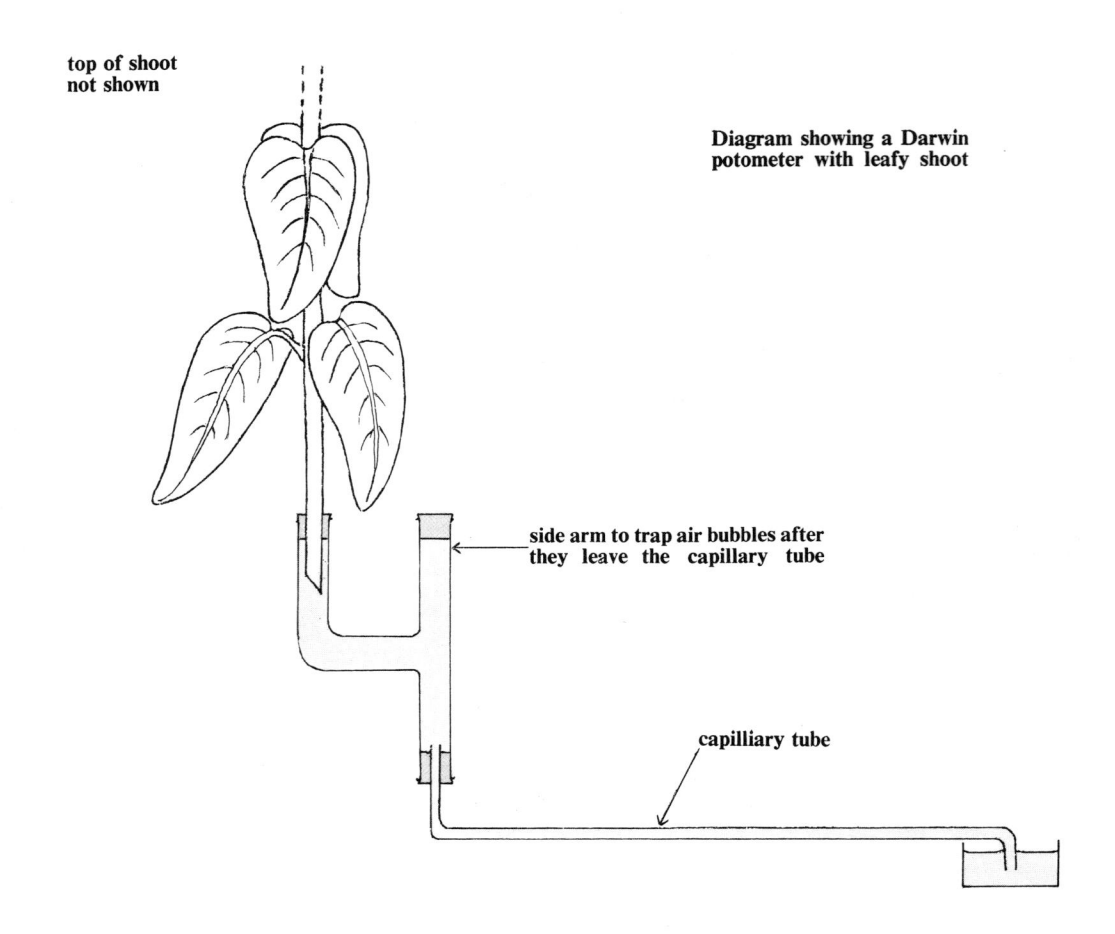

top of shoot not shown

Diagram showing a Darwin potometer with leafy shoot

side arm to trap air bubbles after they leave the capillary tube

capilliary tube

When a reading is to be taken remove the end of the capillary tube from the water in the glass dish and allow a small bubble of air to enter, then replace the end of the capillary tube in the water again. The bubble of air marks a break in the column of water in the capillary tube. As water is taken up by the leafy shoot so the bubble of air moves along the capillary tube.

continued on page 28

Farmer's potometer

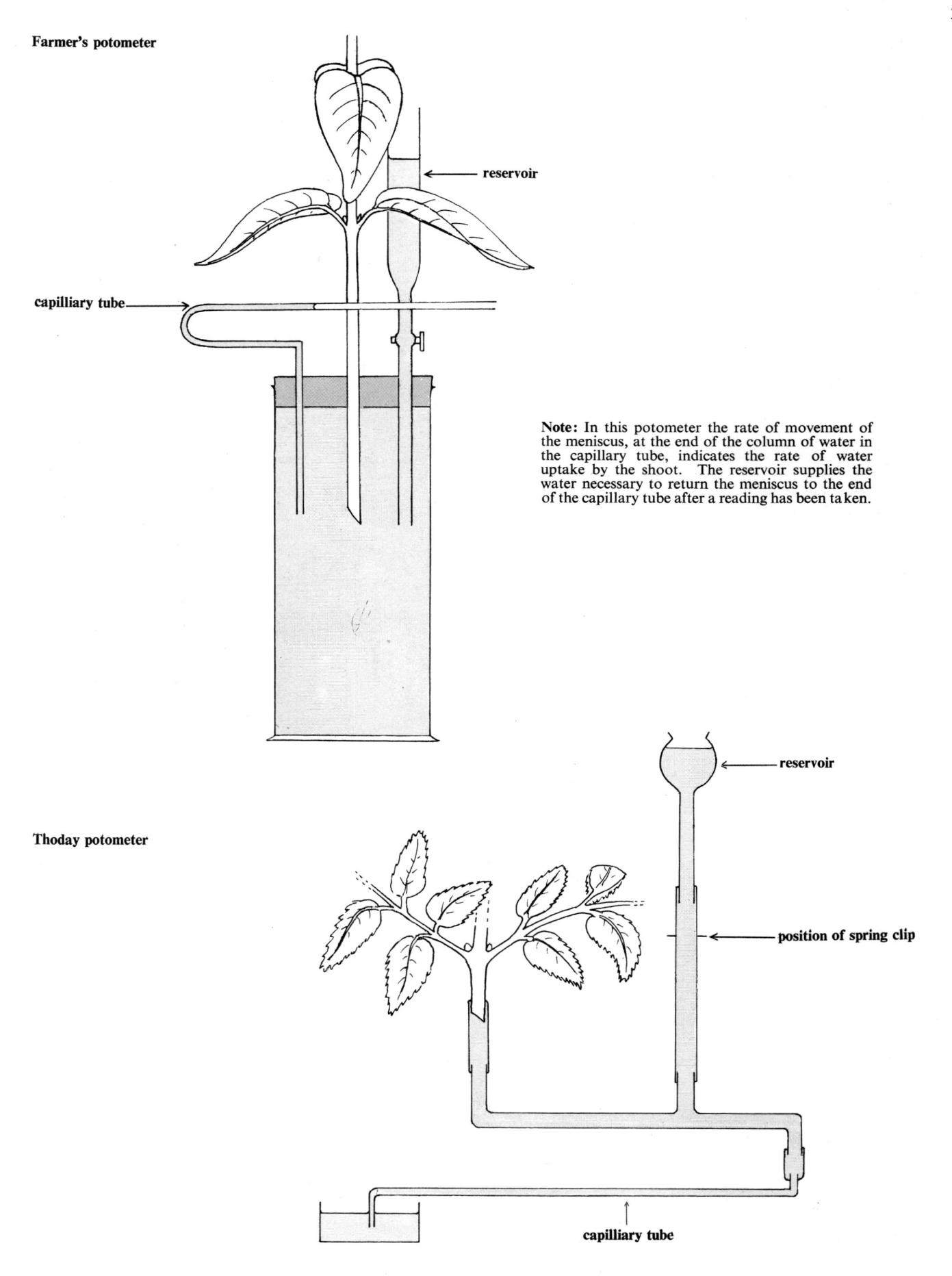

capilliary tube

reservoir

Note: In this potometer the rate of movement of the meniscus, at the end of the column of water in the capillary tube, indicates the rate of water uptake by the shoot. The reservoir supplies the water necessary to return the meniscus to the end of the capillary tube after a reading has been taken.

Thoday potometer

reservoir

position of spring clip

capilliary tube

c

continued from page 26

It is possible to measure either the time taken by the bubble to move a set distance along the capillary tube or the distance which the bubble moves in a given time. Whichever method is used the results will give the amount of water taken up by the shoot in a given time, and therefore the amount of water transpired by the shoot in a given time. On some potometers it is possible to read the volume of water directly, with others it has to be calculated knowing the distance travelled and the internal diameter (bore) of the capillary tube.

If the apparatus is not fitted with a scale then a ruler, fixed in position behind the capillary tube using rubber bands, will suffice.

Some potometers are fitted with a reservoir and after taking a reading, the bubble of air may be pushed back to the open end of the capillary tube, by opening the tap of the reservoir and allowing water to enter the apparatus.

Take several measurements. Then the apparatus may be moved to a position in which the atmospheric conditions contrast as much as possible with the original conditions. It should be left for 15 minutes to become adjusted to its new surroundings, after which further readings of uptake of water may be recorded. This procedure may be repeated several times, it is then possible to compare the rate of transpiration under different conditions

After determining the volume of water it is necessary to determine the total area of leaf surface. To do this remove all the leaves, cut at random from the leaves squares of side 1cm. Weigh these squares and the average weight is taken to be the weight of one square centimetre of leaf surface. Although the square has two surfaces it is taken that only one of these contains stomata and is therefore transpiring. To calculate the total area of leaf surface which is transpiring, weigh the cut squares again, together with all the leaves. Divide the total weight by that calculated to be the weight of a centimetre square to determine the total area of leaf surface.

The presence of stomata on only one surface of the leaf may be verified by microscopic examination.

Specimen calculation:—

weight of **n** centimetre squares of leaf = **w** grams

∴ average weight of 1 centimetre square = $\frac{w}{n}$ grams

let the total weight of the leaves = **T** grams

then the total area of transpiring leaf surface = $\frac{T}{w} \times n$ = **A** cm²

let the distance moved by air bubble along capillary tube = **L** mm

the time taken to move this distance = **t** seconds

the diameter of capillary tube = **d** mm

volume of water taken up by shoot = $\frac{\pi d^2 L}{4}$ = **V** mm³

in **t** seconds this volume was transpired by **A** cm² of leaf surface

in **1** second the volume of water transpired = $\frac{V}{tA}$ mm³/cm² of leaf surface

∴ volume of water transpired = $\frac{V}{1000tA}$ ml/cm²s

Experiment 24
To compare the functioning of different types of potometer by calculation of the standard error of each

APPARATUS REQUIRED
potometers
leafy shoots

METHOD OF PROCEDURE
The potometer should be set up under water as described in experiment 23 on page 26. Remove the potometer from the water and place it in a warm well aerated position and leave it for 15 minutes to adjust itself to its surroundings.

Take about 10 separate readings of the distance travelled along the capillary tube by a bubble of air in 2 minutes. The readings will probably vary slightly from each other due to experimental error.

Each potometer has its own experimental factor called the standard error which may be calculated from the following equation:—

$$\text{Standard error} = \frac{\xi d^2}{n(n-1)}$$

Where:— **d** = the deviation of the results from the arithmetic mean of the results. It is determined by calculating the arithmetic mean and then determining the deviation of each result from this mean
ξd^2 = the sum of the squares of the deviations
n = the number of observations

This procedure should be repeated with the other potometers, making sure that conditions, especially temperature, are identical in each case. Then the standard error may be calculated as above for each potometer.

A comparison can be made of the accuracy of the functioning of the potometers by calculating the % of experimental error for each potometer. That is the % of the standard error of the arithmetic mean.

Experiment 25
To demonstrate leaf suction

APPARATUS REQUIRED
fresh leafy shoot
piece of wide glass tubing
2 rubber bungs to fit wide glass tubing each with a single hole through it
piece of narrow glass tubing to fit the hole in one of the bungs
mercury
dish with vertical sides.
clamp and stand

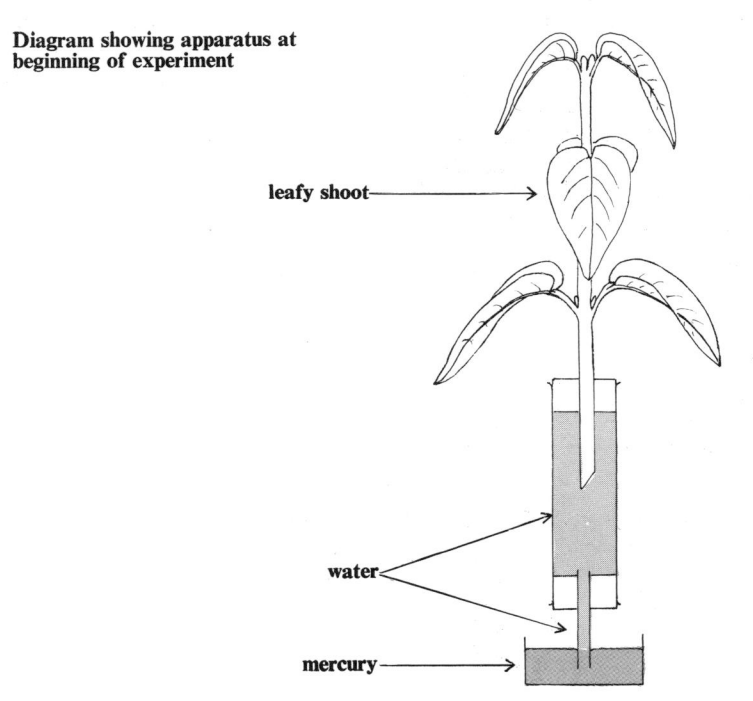

Diagram showing apparatus at beginning of experiment

leafy shoot
water
mercury

METHOD OF PROCEDURE
Fill a deep sink or bowl with water. Pass one end of the narrow glass tubing through the hole in one of the bungs. Immerse the wide glass tubing, bungs and narrow glass tubing in the water in the sink. Make sure that the pieces of tubing are full of water. Immerse the cut end of the fresh leafy shoot in the water and remove about 1 inch of the stem, at the cut end. Pass the cut end of the stem through the hole in the second bung. Insert the bungs, one in each end of the wide glass tubing making sure that no air bubbles are trapped in the apparatus, and keeping the leaves of the shoot as dry as possible. Place a finger over the open end of the narrow glass tubing then remove the apparatus from the sink. Holding the tubes in a vertical position immerse the end of the narrow glass tubing in a dish of mercury. Do not remove finger until the end of the tube is below the surface of the mercury. Support the apparatus using a clamp and stand. Leave for several hours, note any rise of mercury up the glass tubing, which will demonstrate that the suction within the leaf is adequate to raise a column of mercury.

Experiment 26
To demonstrate root pressure

APPARATUS REQUIRED
potted plant
rubber pressure tubing, the diameter of the bore being such that it will fit tightly around the stem of the plant
straight piece of narrow bore glass tubing, to fit tightly into the pressure tubing
oil with dropping pipette
clamp and stand

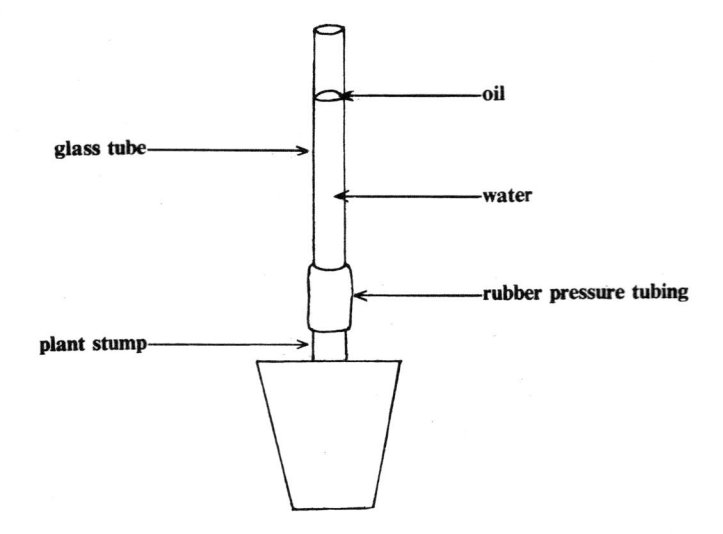

METHOD OF PROCEDURE
Fill a deep sink or bowl with water. Place the pot with plant in the water so that most of the shoot is immersed. Cut the stem of the plant, under water, about 2 inches from the surface of the soil in the pot and remove the shoot. Slip the piece of rubber pressure tubing over the end of the stump of the plant. The diameter of the tubing should be such that it fits tightly around the stump. Immerse the straight piece of narrow bore glass tubing in the sink of water tilting it so that the lower end is full of water. Insert this into the open end of the rubber tubing.

Remove the pot from the sink and support the glass tubing by means of a clamp and stand. Place a drop of oil on the surface of the water in the glass tubing; the level of the water in the tubing may first drop and then become steady. Mark the position of the water surface when it ceases to drop.

Leave the apparatus for 2-3 days noting any change in the level of the water in the glass tubing. Make sure that the soil in the pot is kept moist throughout the experiment.

Experiment 27
To demonstrate the opening and closing mechanism of a stoma

APPARATUS REQUIRED
Zebrina plant which has been exposed to bright light for several hours previously
microscope and glass slides
micrometer eyepiece
filter paper
sucrose solution, approximately 1M
water

METHOD OF PROCEDURE
Remove a leaf from the Zebrina plant. Strip off a piece of the lower epidermis and mount it in water on a slide under a coverslip. Place the slide on the microscope stage, then using the high power objective determine which stoma has the widest opening. Determine its width using the micrometer eyepiece.

Place 2-3 drops of the sucrose solution against one side of the coverslip and draw it underneath using a piece of filter paper placed against the opposite side of the coverslip. Observe the stoma for a few minutes noting any change which takes place.

Replace the sucrose solution with water using filter paper as above. Repeat the "washing" with water several times to make sure that all of the sucrose solution is removed. Place the microscope in a position in which the slide is exposed to bright light. Observe from time to time noting the time taken for the opening of the stomatal pore to become noticeable.

PHOTOSYNTHESIS

In order to live a plant needs various food substances, water and minerals and a continual supply of energy for use in its vital activities. The majority of plants synthesise their own food substances, obtaining the raw materials, which are carbon dioxide and water, from their surroundings. It is also necessary for the plant to be in general good health for it to synthesise food in adequate quantities. If it is diseased or stunted in any way this obviously affects its normal processes, including food synthesis.

The method by means of which plants synthesise their food substances is called **photosynthesis**. In most cases the initial end product is carbohydrate in the form of a hexose sugar, which is easily converted into the various other types of food substances, as required by the plant, by the wide variety of enzymes which are present in plants.

The final products of synthesis vary depending on the age and type of the plant. For example a young plant produces more proteins and less carbohydrate than an older plant. In some algae, such as Vaucheria, the end product is fat or oil which is stored in globules in the cytoplasm; and a group of brown algae produce the polysaccharide laminarin instead of starch.

Source of energy in a plant cell

The plant cell gets the energy it requires for all its vital activities from **adenosine triphosphate** which may be represented by the letters **ATP** for brevity and convenience. Some ATP is made in the mitochondria from **ADP, adenosine diphosphate,** and some is made during the light stage of photosynthesis (see page 34). The mitochondria are involved in the oxidation of various organic molecules, during which oxidations energy is released and is stored in high energy bonds in the ATP. The adenosine triphosphate molecule contains three phosphate groups, while, as its name implies, the adenosine diphosphate molecule contains only two phosphate groups. In ADP one phosphate group is attached by a bond of normal energy level while the other is attached by a high energy level bond. In ATP one phosphate group is attached by a normal energy level bond as in ADP, while the other two are attached by high energy level bonds. When the third or terminal phosphate of ATP is removed (ATP being converted to ADP), then the high energy bond by means of which the terminal phosphate is attached is broken and the energy stored within it is released.

Diagram of ATP molecule

Energy can be stored in the ATP until it is needed and can be transported to wherever it is required.

Two other substances which are involved in energy transformations within the cell are **nicotinamide adenine dinucleotide phosphate (NADP and NADPH)** formerly referred to as **triphospho-pyridine nucleotide (TPN and TPNH)**, and **nicotinamide adenine dinucleotide (NAD and NADH)** formerly called **diphospho-pyridine nucleotide (DPN and DPNH)**. As each is a powerful biological reductant it can readily force its hydrogen atoms onto other molecules. They take part in many oxidation-reduction reactions in living cells one of which provides the energy for phosphorylation, thus some is oxidised to make ATP.

Raw materials and limiting factors for photosynthesis

The raw materials for photosynthesis are carbon dioxide and water. The method by means of which the plant obtains these raw materials is dealt with on page 33. The presence of these raw materials, in adequate quantities, is essential for photosynthesis to take place, but there are certain other factors which may have a limiting effect. They are the presence of light, chlorophyll, and various minerals in minute concentrations.

THE EFFECT OF LIGHT INTENSITY ON PHOTOSYNTHESIS

It has been found that the rate of photosynthesis is directly related to light intensity, other factors such as temperature, availability of carbon dioxide and water etc., being favourable.

Within certain limits an increase in light intensity results in an increase in the rate of photosynthesis, but a point is reached beyond which an increase in light intensity causes no further increase in the rate of photosynthesis. Below this point light intensity is the factor limiting the rate of photosynthesis. Above this point some other factor such as temperature or carbon dioxide concentration becomes the limiting factor.

Graph representing the relationship between the rate of photosynthesis and light intensity

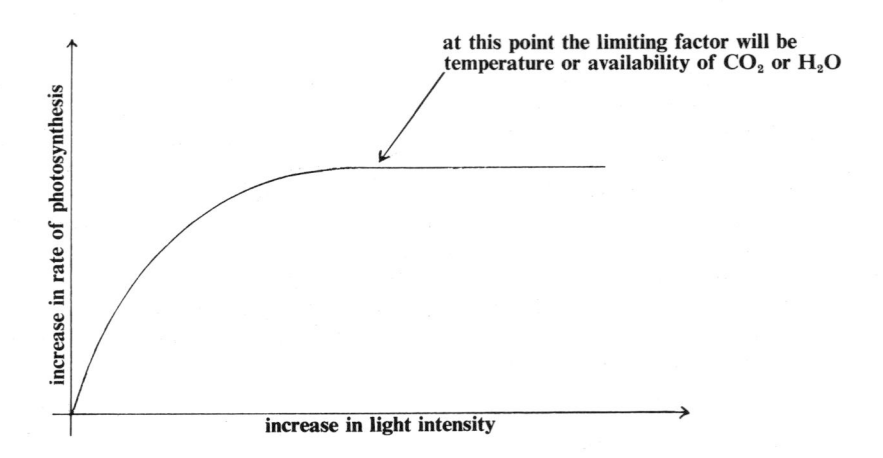

THE EFFECT OF TEMPERATURE ON THE RATE OF PHOTOSYNTHESIS

The optimum temperature for photosynthesis is about 28°C. It will not take place below 0°C nor at temperatures above 40°C.

In conditions of low light intensity an increase in temperature has very little effect on the rate of photosynthesis. At higher light intensities an increase of temperature of for example 10°C, from 18°C to 28°C will result in the doubling of the rate of photosynthesis.

MINERAL REQUIREMENTS CONNECTED WITH PHOTOSYNTHESIS

The presence of certain minerals is also essential for photosynthesis. Magnesium is a component of the chlorophyll molecule and is therefore required for its formation while iron and manganese catalyse the reactions involved in the formation of chlorophyll. Copper and zinc are also probably required in traces. The plant obtains these minerals from the soil (see page 9 for further details).

Photosynthetic pigments

In some parts of the plant the cells contain a green pigment called **chlorophyll**. The pigment is found in bodies called **chloroplasts** and in which two orange-red pigments are also found. The chloroplasts are the site for photosynthesis. They contain the two chlorophyll pigments, **chlorophyll a** and **chlorophyll b,** which vary slightly in their formulae. It is believed that in living cells chlorophyll a exists in two forms which have different light absorbing characteristics. Since only one form is obtained when chlorophyll a is extracted from cells it is assumed that in living cells the two forms differ in the way in which the moleclues of chlorophyll are grouped together, or in the way they are associated with other chemical partners, (Rabinowitch and Govindjee 1965).

formula of chlorophyll a $C_{55} H_{72} O_5 N_4 Mg$

formula of chlorophyll b $C_{55} H_{70} O_6 N_4 Mg$

The two other pigments, **carotene** and **xanthophyll,** which are also present are referred to as **accessory pigments.** They cannot photosynthesise but they assist by absorbing light of wavelengths which cannot be absorbed by the chlorophyll and passing the energy obtained from this on to the chlorophylls which can use it in photosynthesis.

The chlorophyll can only absorb light of certain wavelengths and the presence of pigments of more than one colour ensures that, as much of the sunlight as possible, is utilised. In fact all parts of sunlight could be used as a source of energy for photosynthesis providing compounds capable of absorbing them were present.

Various algae contain, in addition to chlorophyll, certain other pigments which tend to mask the green colour of the chlorophylls, giving the algae their own characteristic colour. Amongst these are the red algae which contain the red pigment **phycoerythrin,** which is capable of absorbing light of short wavelengths. These algae are submerged in sea water all the time and since light of long wavelength tends to be absorbed by water they receive mainly light of short wavelength. If they are to survive under these conditions, it is essential that they should contain a pigment, namely phycoerythrin, which is capable of absorbing this light and passing the light energy on to the chlorophylls.

Photosynthetic tissue, its structure, the way in which it obtains the raw materials for photosynthesis and the removal of the products of photosynthesis

As stated earlier there are special cells in which the photosynthetic pigments are located. These cells are grouped together to form the photosynthetic tissue. The chief photosynthetic organs of the plant are the leaves but they are not the only photosynthetic parts, green stems and stipules are also capable of photosynthesis. Leaves are commonly flattened, the lamina is expanded laterally from a single midrib or a group of large veins, except of course in those cases where they are modified due to special environmental conditions. They provide as large a surface area as possible for absorption of light and gaseous exchange. Leaves are also specialised in that they have an extensive air space system from which the individual photosynthetic cells can draw the carbon dioxide, which they need for photosynthesis, and into which they can discharge the oxygen formed during this process. The air space system is in connection with the external atmosphere through the stomata. These are minute pores in the epidermis of the leaf each of which is surrounded by a pair of guard cells which regulate the size of the pore, (see Book 1 page 31). The gases pass through the stomatal pores and internal air spaces of the leaf by diffusion. If the photosynthetic cells are continually removing carbon dioxide from the surrounding air spaces, then the concentration of the carbon dioxide near the cells is lower than that in other regions, especially near the stomatal pores, and carbon dioxide diffuses towards the air spaces near the photosynthetic cells along a concentration gradient from a higher concentration in the region of the sub-stomatal air cavities, to a region of lower concentration near the photosynthetic cells. Similarly, since the photosynthetic cells discharge oxygen produced as a by-product of photosynthesis into the adjacent air spaces, the concentration of oxygen near these cells rises above that in other regions and oxygen diffuses away from the air spaces near the photosynthetic cells towards the sub-stomatal air cavities, along a gradient of decreasing concentration.

Details of the lamina of Prunus leaf from a transverse section

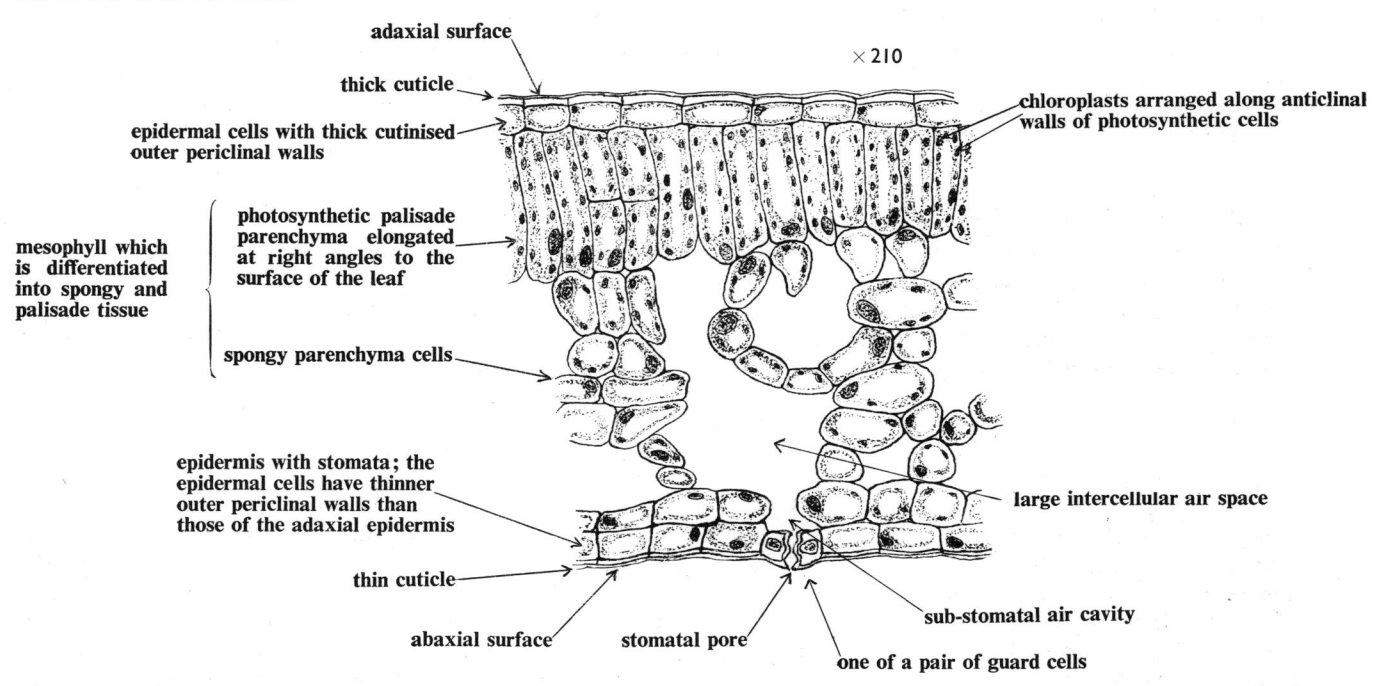

Note: vascular tissue not included

In daylight when respiration and photosynthesis are proceeding simultaneously then some of the oxygen which is produced as a by-product in photosynthesis may be used in the respiratory processes; while the carbon dioxide produced in respiration may be used for photosynthesis. The result being that a plant which is in the light and obviously photosynthesising as well as respiring takes carbon dioxide from, and loses oxygen to, the surrounding atmosphere, the amounts depending on the relative rates of photosynthesis and respiration. This movement of gases between the external atmosphere and the air space system within the leaf is via the stomatal pores. The gases diffuse from a region of higher concentration to a region of lower concentration and in this way oxygen passes to the external atmosphere while carbon dioxide passes inwards.

There is an extensive conducting system consisting of a network of minute veins passing through the leaf in all directions so that no cell is far from one. Each vein consists of a vascular tissue strand which is often surrounded by a sheath of parenchyma or fibres. The photosynthetic cells obtain the water they require from the xylem of the veins. The water enters the plant through the root hairs and is passed across the root to the central xylem. It then moves upwards through the plant, in the xylem, and passes into the system of veins within the leaves (see pages 6-8). The veins also provide a very efficient system for the removal of the food products of photosynthesis, from their source in the photosynthetic cells to various other parts of the plant where they may be required or may be stored; thus preventing a build up of food substances in the photosynthetic cells which would hinder further production.

Although the removal of food products is very efficient, it is possible for sugars to be formed in quantities in excess of the conducting capacity of the phloem. The accumulation of these sugars in large quantities could upset the leaf metabolism, this is prevented by the conversion of some of the sugar to starch.

MECHANISM OF PHOTOSYNTHESIS

The process of photosynthesis is usually divided into two stages.

The **first stage** involves two processes:—

1. The removal of hydrogen atoms from water and the resulting production of oxygen molecules. This is the process which at present is least understood. It is known to consist of a series of reactions which probably require several enzymes, one of which contains manganese. This process is often referred to as the **photolysis of water.**

$$H_2O + \text{electron acceptor } Z \rightarrow ZH + O_2$$

2. The second process involves the absorption of energy from sunlight and its use in the transference of the hydrogen atoms removed from water in the first process of this stage to the second stage or dark stage. This process is commonly referred to as the **light stage,** see below.

The **second stage** is often referred to as the **dark stage** since it does not require the presence of light. It is during this stage that the carbohydrate is synthesised using carbon dioxide and the end products of the light stage. This stage is continuing at the same time as the first stage but runs at a lower rate, so the end products of the first stage tend to accumulate. The dark stage can continue in the absence of light, until the accumulated end products are used up.

Details of the light stage of photosynthesis

The presence of light is essential for the processes involved in this stage to proceed. It involves the absorption of energy from sunlight by certain pigment compounds and the use of this energy in a series of reactions which transfer the hydrogen atoms or electrons, removed from water, to the second or dark stage, in which the hydrogen atoms are used to reduce carbon dioxide to carbohydrate. At the same time some of the energy trapped by the pigments is stored in high energy bonds in ATP as the result of a phosphorylating reaction.

$$\underset{\substack{\text{(two phosphate} \\ \text{compound)}}}{ADP} + \text{phosphate molecule} + \text{energy} \rightarrow \underset{\substack{\text{(three phosphate} \\ \text{compound)}}}{ATP}$$

From the work done by E. I. Rabinowitch and Govindjee, the late Robert Emerson and co-workers, Robert Hill, Fay Bendall, Duysens and his associates and other workers it is now believed that cells possess two light absorbing systems. System one containing one form of chlorophyll a, which absorbs light of longer wavelengths mainly around and above 680 millimicrons and the other referred to as system two containing the other form of chlorophyll a, absorbing light of about 670 millimicrons. The chlorophyll a in system two appears to be assisted by chlorophyll b, acting as an accessory pigment, and various other accessory pigments found in brown, red and blue-green algae. It has been suggested however that the distribution of pigments is not as clear cut as this, since in the red algae, a large amount of the chlorophyll a, absorbing at around 670 millimicrons, seems to belong to system one rather than system two. It would appear that the two systems provide energy for two different photochemical reactions and for efficient photosynthesis the rates of the two reactions must be the same. Important in these reactions are the **cytochromes**. **Cytochromes** are proteins that have a chemical group attached to them which contains an iron molecule. They are found in mitochondria and are involved in the reactions in respiration. It is known that chloroplasts contain cytochromes of two kinds one called cytochrome f, the other cytochrome b_6.

In 1960 Robert Hill and Fay Bendall proposed a hypothesis concerning the action of these cytochromes as carriers of electrons connecting the two photochemical systems. They suggested that an electron is passed to the cytochrome b_6 in system two, from the electron donor ZH produced at the end of the first process. This electron is then passed on to cytochrome f, in system one, by a reaction which does not require light energy since the oxidation reduction potential of cytochrome f, is more positive than that of cytochrome b_6. During this process energy is released and stored in ATP during a phosphorylating reaction when ADP is converted to ATP. The electron is passed from cytochrome f, to the acceptor X in the third stage of photosynthesis, the energy of the electron being increased during the process by light energy absorbed by the pigments of system one.

Thus energy is absorbed by the chlorophyll during the photochemical reactions. Part of this energy is released in the reaction between the two cytochromes and most of this is stored in ATP. The rest of the energy is passed on to the third stage, in the form of high energy electrons or hydrogen atoms, which are passed to the acceptor X to be used in the reduction of carbon dioxide to carbohydrate.

DIAGRAM REPRESENTING THE MOVEMENT OF ELECTRONS IN THE LIGHT STAGE OF PHOTOSYNTHESIS

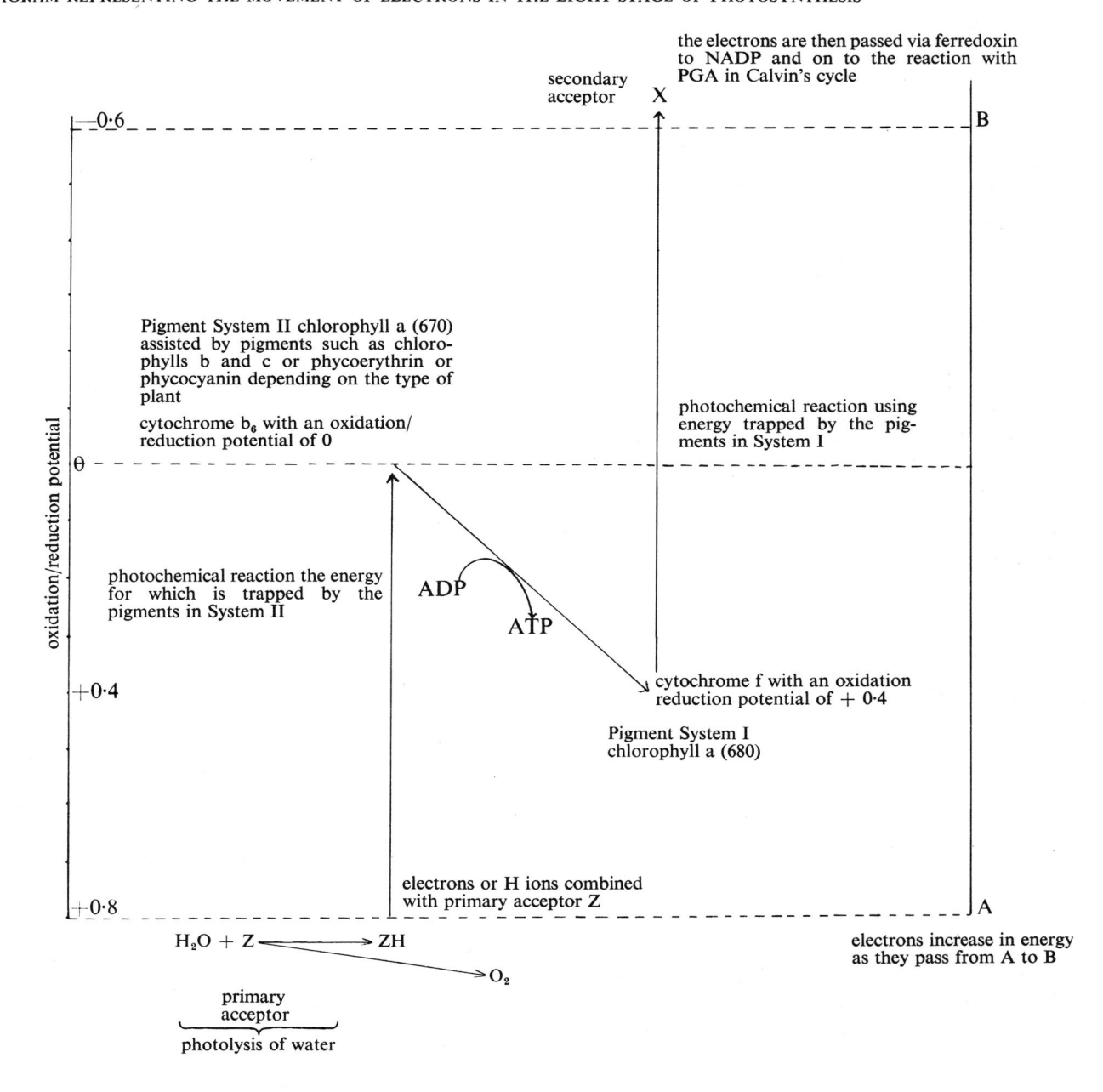

DIAGRAM ILLUSTRATING THE CYCLIC SERIES OF PROCESSES FOLLOWED BY CALVIN'S SCHEME

One molecule is removed from the system with each cycle. It may be removed here in the form of PGA or at some other point in the cycle depending on the final end product required.

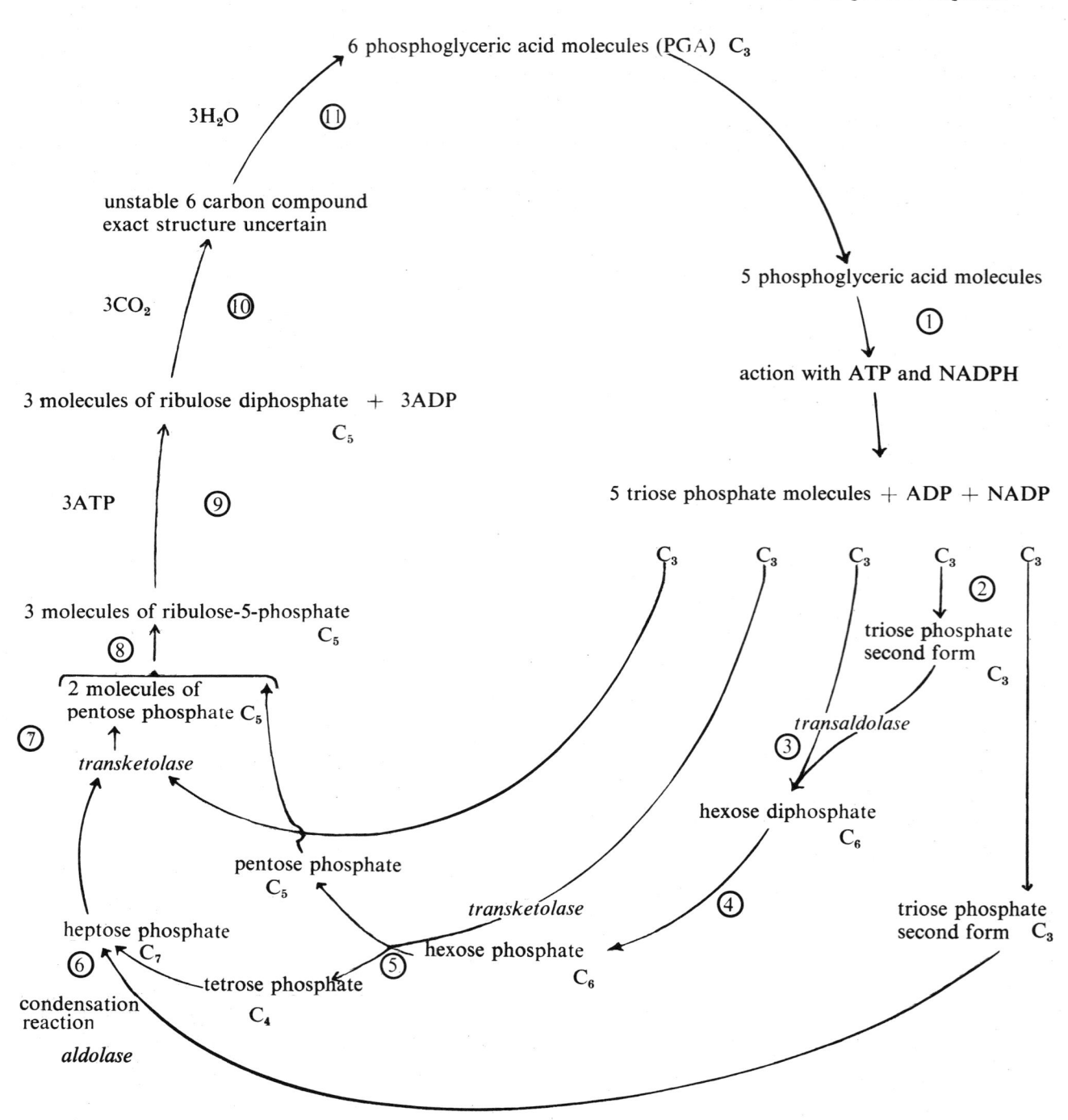

6 phosphoglyceric acid molecules (PGA) C_3

$3H_2O$ ⑪

unstable 6 carbon compound exact structure uncertain

$3CO_2$ ⑩

3 molecules of ribulose diphosphate + 3ADP C_5

3ATP ⑨

3 molecules of ribulose-5-phosphate C_5

⑧

2 molecules of pentose phosphate C_5

⑦ *transketolase*

heptose phosphate C_7

⑥ condensation reaction

aldolase

tetrose phosphate C_4

⑤ hexose phosphate C_6

transketolase

pentose phosphate C_5

5 phosphoglyceric acid molecules

① action with ATP and NADPH

5 triose phosphate molecules + ADP + NADP

C_3 C_3 C_3 C_3 C_3

② triose phosphate second form C_3

transaldolase ③

hexose diphosphate C_6

④

triose phosphate second form C_3

Note: the numbers in the circles refer to the corresponding stages in the descriptive text on page 37.

Details of the dark stage of photosynthesis

This uses carbon dioxide and the energy stored in ATP during the light stage, also the high energy electrons or hydrogen atoms which have been passed to the hydrogen acceptor X. It results in the synthesis of carbohydrate which is easily converted into other food substances such as fats and proteins. The final end product varies depending on the plant and its age (see page 31). This is the limiting stage.

The most recent scheme and that most in favour at present, was put forward by Melvin Calvin and is referred to as **Calvin's scheme**. It describes a cyclic process which involves various sugars from trioses to heptoses. In this cycle five molecules of **phosphoglyceric acid** (PGA for brevity and convenience), are passed through a series of complex reactions, during which energy obtained from ATP and NADPH is used, and carbon dioxide is combined with the sugars with the resulting formation of six PGA molecules. There is a net gain of one molecule of PGA per cycle.

The additional material synthesised in this process may be removed at various points according to the final end product required. If this is to be fatty acids, fats, amino acids or carboxylic acids it is removed in the form of PGA. If the end product is to be starch or sucrose then it is removed after the conversion of PGA to triose phosphate and thence to hexose phosphate. Fats may also be formed from triose phosphate via glycerol.

The high energy electrons or hydrogen atoms from the light stage combine with the acceptor X and are passed on to ferredoxin (Fd for convenience) and then on to NADP.

The PGA molecules are reduced by ATP and NADPH to form five triose phosphate molecules ①. Two of these triose phosphate molecules are converted into another form of triose phosphate ②, one of which then combines, in a condensation reaction, with one of the other triose phosphates to form a hexose diphosphate ③. This loses one of its phosphate groups to form hexose phosphate ④, which in the presence of the enzyme transketolase reacts with one of the original triose phosphate molecules to form a tetrose and a pentose ⑤. The tetrose reacts with a molecule of triose phosphate which has been converted into the second form, and in a condensation reaction, catalysed by the enzyme adolase, forms a heptose phosphate ⑥. This heptose phosphate then reacts with the last of the original triose phosphates, in the presence of the enzyme transketolase, to form two molecules of pentose phosphate ⑦. In all, three molecules of a pentose phosphate have been formed. These are converted into ribulose-5-phosphate ⑧, then to ribulose diphosphate ⑨, the phosphate groups coming from ATP, which is at the same time converted to ADP. The three carbon dioxide molecules are then added to the ribulose diphosphate molecules to form three molecules of an unstable six carbon substance the structure of which is unknown ⑩. These react with three water molecules, with the resulting formation of six PGA molecules ⑪, a net gain of one PGA molecule.

PHOTOSYNTHESIS EXPERIMENTS

Experiment 28

To demonstrate the presence of starch in a green leaf

APPARATUS REQUIRED

green leaf which has been exposed to light for about 12 hours previously, and removed from the plant just before the experiment (geraniums are useful for starch tests on leaves and are easily obtained and kept in the laboratory greenhouse)
2 beakers
water, preferably hot
2 bunsen burners, tripods and gauzes
glass boiling tube
white porcelain tile
shallow dish
solution of iodine in potassium iodide, 0·3 g iodine and 1·5 g potassium iodide in 100 ml water

METHOD OF PROCEDURE

Half fill each beaker with hot water then place each on a tripod, fitted with a gauze, and heat with the bunsen burner until the water is boiling. Turn both burners out. This is important since the vapour from ethyl alcohol, which is to be used now, is inflammable.

Kill the leaf by placing it in one of the beakers of boiling water and leaving for a few minutes. Half fill the boiling tube with ethyl alcohol, and place it in the water in the second beaker which should have cooled slightly by now. Remove the leaf from the water in the other beaker and place it in the boiling tube with the ethyl alcohol.

The leaf will gradually loose its colour since the chlorophyll is dissolved out of the dead leaf tissue by the alcohol. The actual colour varies depending on the thickness and the type of leaf. When the leaf has lost its green pigment remove it from the boiling tube, wash it in boiling water and place it in a shallow dish containing dilute iodine solution. After a few minutes remove and wash the leaf, then place it on a white tile and note any colour change.

Note: Starch stains a very dark blue with iodine in potassium iodide solution but other substances stain dark brown.

Experiment 29

To demonstrate that light is necessary for photosynthesis

APPARATUS REQUIRED

potted plant such as a geranium which has been kept in the dark for at least 48 hours immediately preceding the experiment
black paper
transparent adhesive tape or paper clips
apparatus as used in experiment 28 for testing for the presence of starch in a leaf.

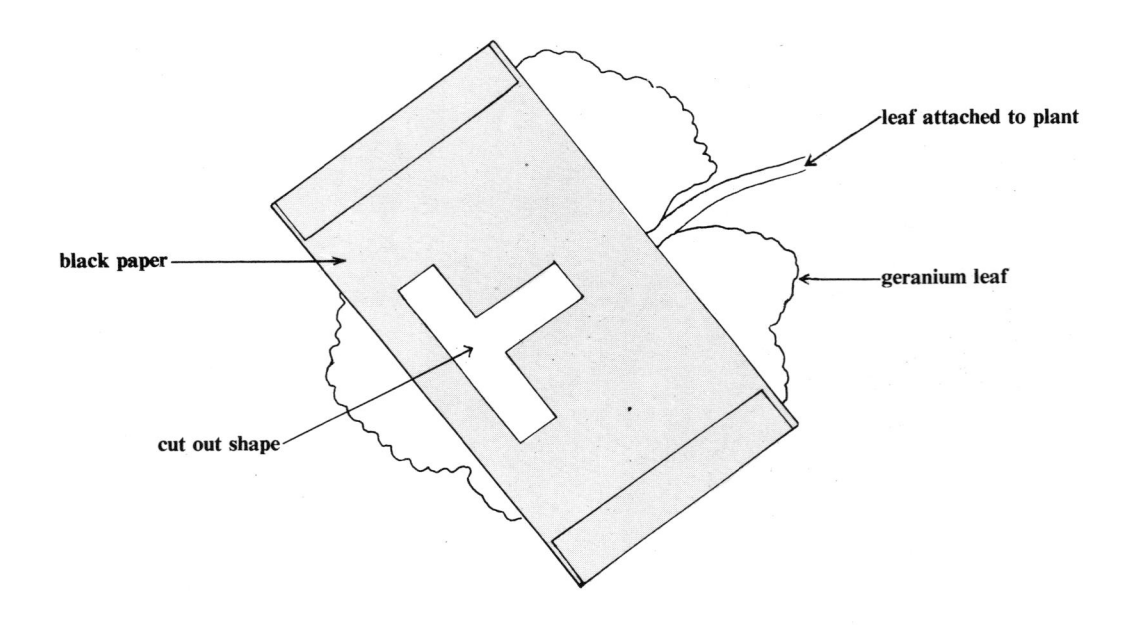

METHOD OF PROCEDURE

Remove a leaf from the plant and test for the presence of starch as in experiment 28 on page 37. If the result is negative then you may proceed with the experiment. (If a positive result is obtained then the plant should be returned to the dark for a further 24 hours after which the experiment should be started again).

Select the leaf to be used, do not remove it from the plant. Take 2 pieces of black paper large enough to cover most of the leaf and to overlap on 2 sides. Cut a shape from the centre of one of the pieces of paper and discard the cut out shape. Place the piece of paper on the adaxial surface of the leaf and place the other intact piece on the abaxial surface of the leaf; fix the 2 pieces of paper together at their adjoining edges by means of the transparent adhesive tape or paper clips, so that light is excluded from the region which they cover, but there is free access of gases to all regions. Place the plant in the light and leave it for about 48 hours, then remove the leaf from the plant and test for the presence of starch as in experiment 28 on page 37. From your results decide whether light is essential for photosynthesis to take place.

Note: Your results should include a diagram of the leaf covered with paper and a diagram of the leaf after staining. Do not use a variegated leaf.

Experiment 30
To determine whether chlorophyll is necessary for photosynthesis

APPARATUS REQUIRED

variegated leaf from a plant which has been kept in the light for several hours previously, (a suitable plant would be a variegated geranium).
apparatus as in experiment 28 on page 37 to test for the presence of starch in a leaf.

METHOD OF PROCEDURE

Make a plan of the leaf to show those regions in which chlorophyll is present (that is, those which are coloured green). Test this leaf for the presence of starch as in experiment 28 on page 37. Compare the leaf after it has been removed from iodine with the plan showing the distribution of chlorophyll within it, and determine whether chlorophyll is necessary for photosynthesis.

Note: Your results should include a plan showing distribution of chlorophyll within the leaf and a plan of the leaf after staining with iodine.

Experiment 31
To demonstrate that carbon dioxide is necessary for photosynthesis

APPARATUS REQUIRED

potted plant which has been kept in the dark for the previous 48 hours
conical flask
white petroleum jelly
soda lime
cork split in half, with a hole hollowed out of the centre just large enough to take the leaf stalk
apparatus to test for the presence of starch as in experiment 28 on page 37.
clamp and stand

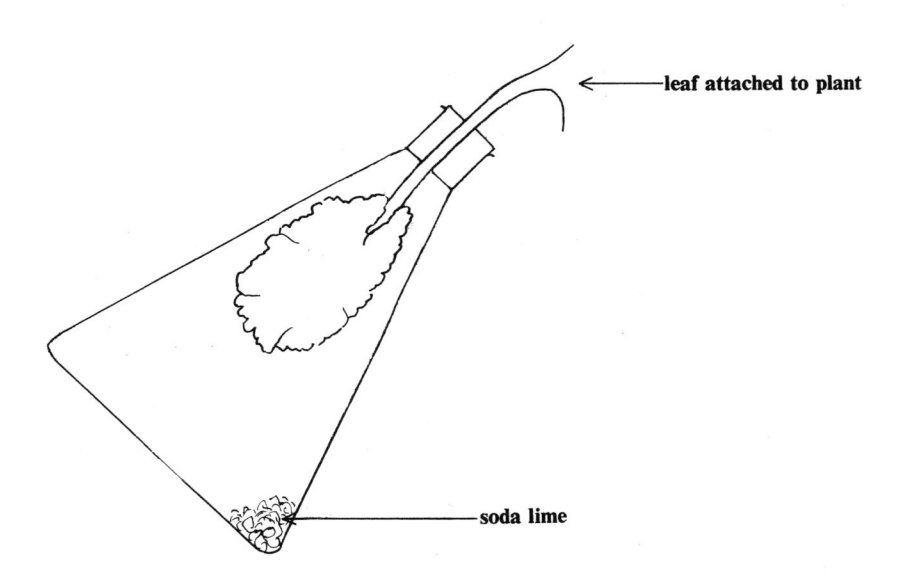

METHOD OF PROCEDURE

Remove a leaf from the plant and test for the presence of starch as in experiment 28 on page 37. If there is a negative result then you may proceed with the experiment. If the result is positive the plant should be returned to the dark for a further 24 hours.

Place some soda lime in the bottom of the flask. Soda lime will remove carbon dioxide from the air in the flask. Select the leaf to be used in the experiment but do not remove it from the plant. Fix the cork around the leaf stalk making sure there is a good airtight fit by surrounding the stalk with white petroleum jelly first. Insert the leaf into the flask until the cork which surrounds the stalk completely plugs the neck of the flask. Support the flask by a clamp and stand and leave in bright sunlight, or artificial light, for several hours. Then remove the leaf from the flask and test this and another leaf from the plant for starch. From your results decide whether the presence of carbon dioxide is necessary for photosynthesis.

Experiment 32

To determine whether a gas is given off as a result of photosynthesis, to identify it, and to determine whether photosynthesis occurs in the dark.

APPARATUS REQUIRED
3 sets of the following are required:—
   several shoots of Elodea canadensis
   beaker
   water
   test tube
   short stemmed funnel
   clamp and stand

Apparatus at the beginning of the experiment

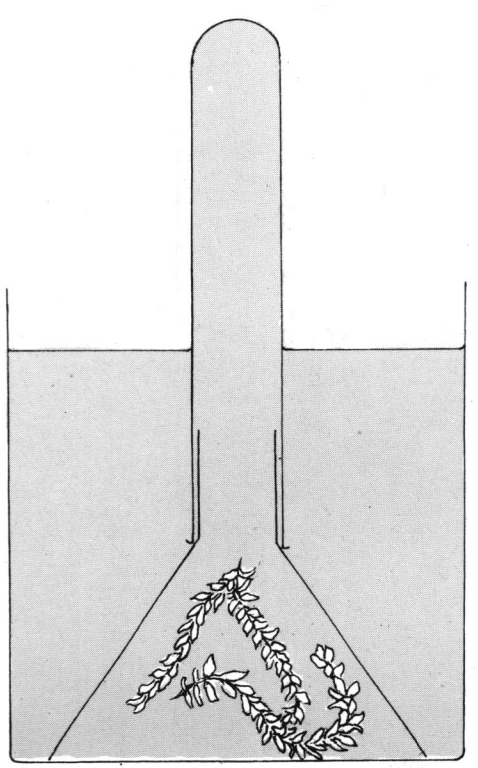

METHOD OF PROCEDURE
Place a few shoots of Elodea in a beaker of water at approximately room temperature. Better and quicker results are obtained if the water has been saturated with carbon dioxide by dissolving one per cent of potassium bicarbonate in it. Place the short stemmed funnel over the Elodea, making sure that the end of the funnel is about 1 inch below the surface of the water in the beaker. Fill a test tube with water, place thumb over the open end and invert the test tube, then lower it into the beaker of water. Remove thumb and place the open end of the test tube over the end of the stem of the funnel, making sure that no water is lost from the test tube, which should then be supported by a clamp and stand. The apparatus should be left in the light for several hours.

  Set up an identical experiment and leave this in the dark for several hours.

  Set up a similar experiment in the light, leaving out the Elodea, this is the control.

  After a short time it will be noticed that small bubbles are given off at the cut ends of the Elodea, in the apparatus in the light. These bubbles begin to collect at the top of the test tube. After several hours sufficient gas should have collected to test. Remove the test tube and test the gas with a glowing splint. Remove the other set of apparatus from the dark and note whether any gas has been produced here, or in the control.

Note: A glowing splint is rekindled in the presence of oxygen.

Experiment 33
To demonstrate chemically that oxygen is produced during photosynthesis using indigo-carmine (Kolkwitz' method)

APPARATUS REQUIRED
test tubes
beaker
clamp and stand
shoot of Elodea canadensis
10% solution of sodium hydrosulphite ($Na_2S_2O_4$)
reduced indigo-carmine solution: this solution is made by dissolving about 0·1 g of indigo-carmine in tap water. This forms a bright blue solution. Then carefully drop by drop add the sodium hydrosulphite solution to the indigo-carmine solution which is reduced to indigo-white by the sodium hydrosulphite: care must be taken to avoid adding an excess of the hydrosulphite. Do not shake or disturb the reduced indigo-white solution.

METHOD OF PROCEDURE
Gently remove 20 ml of the indigo-white solution using a pipette; transfer it to a test tube, placing the tip of the pipette in the bottom of the test tube to reduce the contact of the solution with air. If the indigo-white should start to turn blue then the addition of a drop of hydrosulphite should stop this.

Place a small piece of young Elodea shoot in the test tube with the indigo-white solution. Make sure that the test tube is completely full of liquid, then placing the thumb over the open end invert the test tube over the beaker of indigo-white solution. Support the test tube with the clamp then gently place the apparatus in a bright light. After a few minutes note any change in the colour of the liquid surrounding the Elodea.

Note: Reduced indigo-carmine solution is white but this is very easily oxidised by free oxygen such as that produced by a shoot during photosynthesis. In the oxidised state indigo-carmine solution is blue in colour.

Experiment 34
To compare the rates of photosynthesis under differing conditions

APPARATUS REQUIRED
several sets of the following will be required depending on the number of conditions which you wish to test:—
  beaker containing water
  glass rod
  shoot of Elodea canadensis

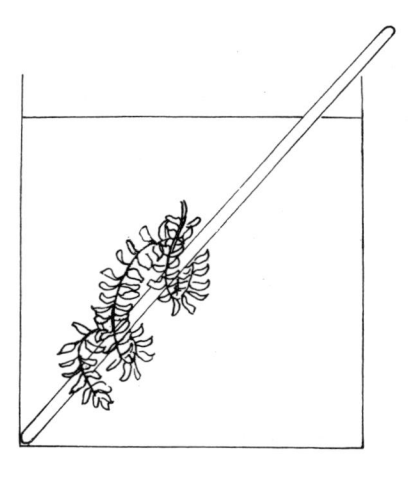

METHOD OF PROCEDURE
Wind a piece of Elodea around each of the glass rods making sure that the cut end is nearest the top of the rod. Place the rods, one in each beaker of water, resting them against one side so that the Elodea is completely immersed in the water as shown in the diagram.

The conditions can be varied by placing one set in bright sunlight, one in the dark and one in medium light, you can also vary the temperature of the water in the beaker having one at room temperature and the others either above or below room temperature.

Leave each set of apparatus to become adjusted to its conditions, after which it will be noticed that bubbles of gas are rising from the cut ends of the Elodea shoots, and that they are rising at different rates in the different sets.

The rates of photosynthesis can be compared under the different conditions by comparing the rates of evolution of the bubbles of oxygen.

Experiment 35

To determine the effects of light and heat on the rate of photosynthesis

APPARATUS REQUIRED
6 sets of the following are required:—
 several shoots of Elodea canadensis
 water
 beaker
 test tube
 short stemmed funnel
 clamp and stand
also required are 3 of each of the following:— water baths, thermometers, tripods and bunsen burners

METHOD OF PROCEDURE
The apparatus should be set up as in experiment 32 page 40, but the conditions in each of the 6 sets should vary slightly, as follows. Three sets in which the water is at room temperature are to be placed in different light intensities, one in the dark, one in moderate light and one in bright sunlight. The other 3 sets are all placed in bright light but the temperature of the water varies, one being at 10°C, another 20°C and the third 30°C, and they should be kept constantly at these temperatures. All 6 experiments should be left for several hours, after which note how much gas is produced in each case. From your results you can decide which conditions favour photosynthesis.

Experiment 36

Ganong's disc method for determining the rate of photosynthesis

APPARATUS REQUIRED
potted plant which has been kept in the dark for the previous 48 hours
sharp cork borer
apparatus to test for starch as in experiment 28 on page 37
desiccator
oven, the temperature of which should be about 110°C

METHOD OF PROCEDURE
Remove a leaf from the plant and test it for the presence of starch. If there is a negative result then you may proceed with the experiment. (If a positive result is obtained the plant should be returned to the dark for a further 24 hours after which it should be re-tested).

Do not remove the leaves from the plant but, using the cork borer, cut a number of discs from one side of several different leaves, trying to avoid damaging any large veins. Then place the plant in surroundings which will favour photosynthesis, that is bright sunlight or if it is a dull day, near a source of bright artificial light, then leave for several hours.

The discs which were removed should then be weighed; more accurate results are obtained if the discs are dried before they are weighed. To dry, place them in an oven at about 110°C for about half an hour then remove them and cool in a desiccator and re-weigh. This should be repeated until the weight is constant.

After several hours cut the same number of discs from the other sides of the leaves using the same cork borer. These discs should then be weighed, and if the first set were dried before weighing then these should also be dried. Compare the weights of the 2 sets of discs and draw your conclusions from any difference that there might be.

Note: You can assume that, had the second set of discs been removed and weighed at the same time as the first set, then the weights would have been identical.

There are variations of this method, for example, Sach's half leaf method where several leaves are split in half, one half being removed in the morning and the other in the evening and the weights compared.

Experiment 37

Extraction and separation of the photosynthetic pigments

APPARATUS REQUIRED
5 g dried nettle powder
90% acetone
92% methyl alcohol
30% potassium hydroxide in methyl alcohol
petroleum ether
distilled water
measuring cylinder
stoppered flask
separating funnel

METHOD OF PROCEDURE FOR THE EXTRACTION OF THE PIGMENTS

Place 5 g of nettle leaf powder (weighed roughly) in a stoppered flask. Add 35 ml of 90% acetone, mix and leave for 15 minutes, during which time the acetone will have taken up most of the pigments from the powder. Filter the extract into a measuring cylinder. Take a separating funnel and place in it a volume of petroleum ether equal to **twice** the volume of the extract in the measuring cylinder. Add the extract to the petroleum ether, insert the stopper in the separating funnel and shake well. The petroleum ether will take up most of the dissolved pigments.

A quantity of distilled water equal in volume to the original extract is then poured slowly down one side of the separating funnel. Care is taken to avoid the formation of an emulsion, swirl the funnel gently but do not shake. After a few minutes the contents of the funnel separate into 2 layers. The lower one, which contains only water and acetone may be run out and discarded, leaving the upper green ether layer. This "washing" procedure should be repeated 3 times, after which the extract of pigments in etherial solution may be run off and used in the following experiments.

METHOD OF PROCEDURE FOR SEPARATION OF THE PIGMENTS USING SEPARATING FUNNELS

Divide a portion of the extract produced as above into 2 parts, and treat as follows.

PART 1

Place the portion of pigment extract in a separating funnel then add 92% methyl alcohol until 2 layers separate out. The upper etherial layer will contain chlorophyll a and carotin, the lower alcoholic layer will contain chlorophyll b and xanthophyll.

PART 2

Place the pigment extract in a separating funnel then add 8 ml of 30% potassium hydroxide in methyl alcohol. Mix well and leave to stand until the green colouration disappears. Add distilled water until the mixture separates into 2 layers. The upper etherial layer contains carotin and xanthophyll. The chlorophylls have been saponified by the alkali. The carotin and xanthophyll may be separated by washing the etherial solution with water in a separating funnel. Run off the water and evaporate the etherial portion down to 1 ml, then dilute with 10 ml of petroleum ether and shake with 10 ml of 90% ethyl alcohol in a separating funnel, leave to stand until 2 layers separate, an upper petroleum ether layer containing carotin and a lower ether layer containing xanthophyll.

Chromatographic separation of the photosynthetic pigments:— see page 91-95

RESPIRATION

Respiration is a term which has more than one application. It may be used with reference to the breathing movements of animals or the exchange of gases between plants and their surroundings. The term respiration is also used with reference to the series of processes occurring within plant and animal cells during which complex food substances are broken down into simpler substances with the simultaneous liberation of energy. It is obvious that these are completely different and it is therefore better to use different terms for the two processes. The term **gaseous exchange** is used when referring to the exchange of gases between the plant or animal and its environment, and the term **tissue respiration** (or **cellular respiration**) when referring to the processes occurring within the plant and animal cells.

Gaseous exchange occurs at the respiratory surface. In the case of a plant this respiratory surface is the leaves, the gases diffusing both inwards and outwards through the stomata, (see Book 1, page 31). Other examples of respiratory surfaces are the lining of the lungs in man and the surface of the gills of fishes. Gaseous exchange is sometimes referred to as external respiration.

As stated above the term **tissue respiration** (or **cellular respiration**) is used with reference to the series of processes occurring within the cells of a plant or animal by means of which complex food substances are broken down into simpler substances, in a stepwise manner, with the ultimate formation of carbon dioxide and water. The important point about this process is that as the food substances are broken down, so the energy stored within them is liberated, and retained by the plant in the form of chemical energy, stored in high energy bonds in certain chemical compounds within the cells.

Energy within the cell

A cell obtains the energy which it requires for its vital activities from **adenosine tri-phosphate** which is represented by the letters **ATP**. ATP is the universal carrier of energy within plants and animals. Each molecule contains three phosphate groups two of which are attached by bonds of high energy level, while the other is attached by a normal energy bond. When the third or terminal phosphate group is removed this high energy bond is broken, the stored energy is liberated and ATP is converted into **adenosine di-phosphate ADP**. The energy is therefore stored in units small enough to be easily transported about the cell and it can be liberated when and where it is required for the vital energy consuming activities of the cell.

The ATP is formed from ADP and phosphate groups in the mitochondria. Research work has established that the mitochondria are the site for tissue respiration, the enzymes involved in the citric acid cycle and the hydrogen transport stages of tissue respiration being located in them; those of the citric acid cycle being in the matrix and those of the hydrogen system in the membrane. The membrane, which is not an inert structure, contracts when ATP concentration is high and relaxes when it is low.

CELLULAR RESPIRATION

In this section when the term respiration is used by itself it will be with reference to tissue respiration. There are two methods by means of which food substances are broken down and the energy stored within them liberated. One process involves the use of oxygen and is referred to as **aerobic respiration**. The other process in which oxygen is not used is referred to as **anaerobic respiration**. The respiratory substrate is the same in both cases (glucose), and they have a common path, that is the series of reactions leading to the formation of pyruvic acid is common to both. See diagram on following page.

The oxidation of one gram molecule of glucose to carbon dioxide and water by burning it in oxygen yields 690,000 calories. According to the fundamental principles of thermodynamics the same amount of energy is always liberated on combustion independent of the process or method; thus the oxidation of one gram molecule of glucose within the cell with the ultimate formation of carbon dioxide and water makes available 690,000 cals of energy, most of which is retained in the form of chemical energy. The oxidation is carried out in a controlled stepwise manner, with the energy being liberated from the glucose derivatives at various stages; this energy being retained by the cells, stored in high energy chemical bonds in the universal energy carrier ATP. As one gram molecule of glucose is oxidised by this process so 38 molecules of ATP are formed, each requiring 12,000 calories of energy for its formation, in all 456,000 calories. Since the oxidation of one gram molecule of glucose yields 690,000 calories, 66% of the energy made available is retained within the plant.

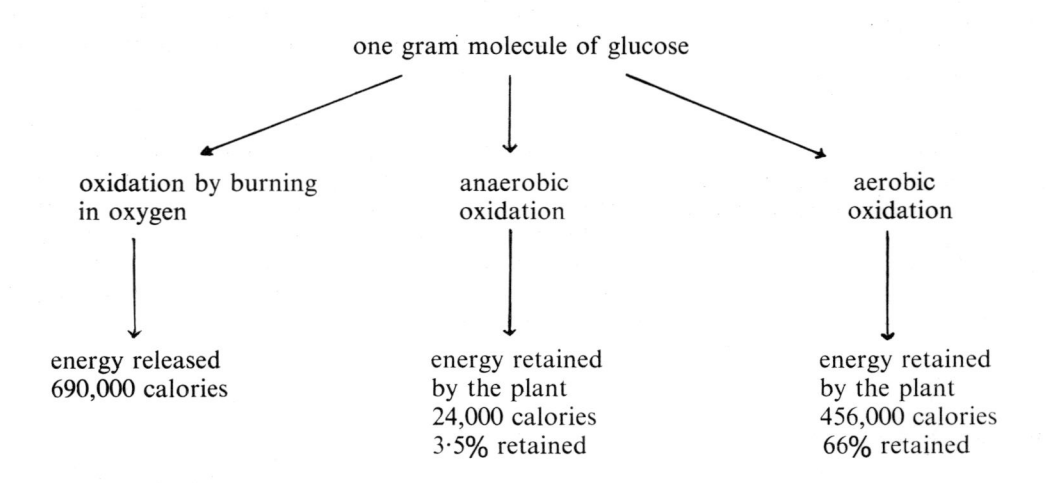

DIAGRAM SHOWING AMOUNTS OF ENERGY RELEASED AND RETAINED BY THE PLANT
AS A RESULT OF THE OXYDATION OF ONE GRAM MOLECULE OF GLUCOSE.

```
                        one gram molecule of glucose
              ↙                      ↓                     ↘
    oxidation by burning       anaerobic              aerobic
       in oxygen               oxidation             oxidation
           ↓                       ↓                     ↓
     energy released         energy retained       energy retained
     690,000 calories         by the plant          by the plant
                              24,000 calories       456,000 calories
                              3·5% retained         66% retained
```

DIAGRAM TO SHOW THE COMMON PATH IN ANAEROBIC AND AEROBIC RESPIRATION

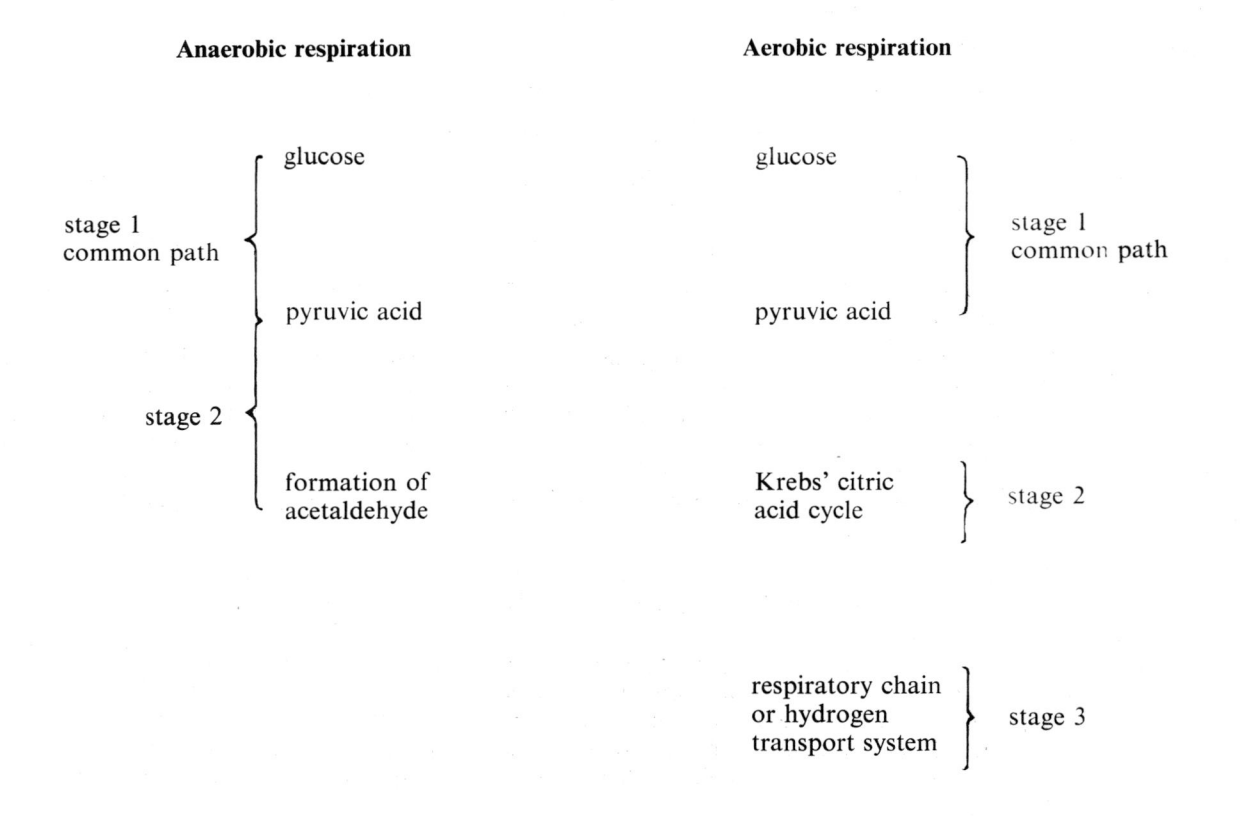

Anaerobic respiration   **Aerobic respiration**

```
                  ┌ glucose                    glucose     ┐
   stage 1        │                                        │  stage 1
   common path   ┤                                         ├  common path
                  │                                        │
                  └ pyruvic acid               pyruvic acid ┘

   stage 2       ┤
                  ┌ formation of               Krebs' citric ┐
                  └ acetaldehyde               acid cycle    ┘  stage 2


                                               respiratory chain ┐
                                               or hydrogen       ├  stage 3
                                               transport system  ┘
```

Aerobic respiration, in which the ultimate end products are carbon dioxide and water, is more efficient than anaerobic in which the end products are carbon dioxide and ethyl alcohol, since break-down continues further in aerobic. As stated on page 44, in aerobic respiration for every gram molecule of glucose broken down there is a net gain of 38 molecules of ATP, but for every gram molecule of glucose broken down anaerobically there is a net gain of only 2 molecules of ATP. In aerobic respiration 66% of the energy available in the substrate is therefore retained by the plant and is available for use in energy consuming activities, while in anaerobic respiration only about 3·5% of the energy available in the substrate is retained, the rest is lost to the plant, most of it remaining in chemical bonds in the end products.

When an organism in which the tissue respiration is normally aerobic is temporarily lacking in oxygen, then tissue respiration can continue for a time anaerobically, although an oxygen debt is built up. An organism in which tissue respiration is normally anaerobic, will not respire aerobically in the presence of oxygen. The reason for this is obvious when considering the details of the two processes (each of which consists of a complex series of reactions involving many enzymes). The first stage of both processes is identical, involving the same enzymes etc., but in aerobic respiration there follow many reactions not occurring in the anaerobic process. A plant which normally respires anaerobically would not possess the enzymes, hydrogen carriers etc., necessary for these reactions and therefore could not make use of any oxygen present.

ANAEROBIC RESPIRATION

The process of anaerobic respiration consists of a series of reactions by means of which glucose is broken down to form ultimately ethyl alcohol and carbon dioxide. Most of the energy liberated during these reactions is retained by the plant, stored in high energy bonds in ATP but most of the energy which was 'potentially available' in the glucose is lost to the plant since it is retained in chemical bonds in the end products.

The various steps involved in this process are summarised below and the reactions, with the enzymes involved, are shown in the table on page 47.

Summary of the processes involved in anaerobic respiration

Stage 1   **The breakdown of glucose to pyruvic acid.**

    Step 1. The phosphorylation of glucose by ATP to form glucose.6.phosphate and ADP.

    Step 2. The conversion of glucose.6.phosphate to fructose.6.phosphate.

    Step 3. The phosphorylation of fructose.6.phosphate to form fructose.1.6.diphosphate, at the expense of a further molecule of ATP which is degraded to ADP.

    Note: So far energy has been put into the system since two molecules of ATP have been used up.

    Step 4. The glycolysis of fructose.1.6.diphosphate to form 3.phospho-glyceric aldehyde (two molecules of the aldehyde being formed from each molecule of the diphosphate).

    Step 5. The conversion of 3.phospho-glyceric aldehyde to 1.3.diphospho-glyceric acid by the enzyme triose phosphate dehydrogenase, which requires the presence of inorganic phosphate and NAD to show activity. This is believed to take place in two stages, the first being the phosphorylation of 3.phospho-glyceric aldehyde to 1.3.diphospho-glyceric aldehyde, with the formation of a high energy bond when the extra phosphate is attached. The second stage is the oxidation of the aldehyde to the acid in the presence of NAD (co-enzyme 1), which acts as a hydrogen acceptor and is therefore reduced.

    Step 6. A trans-phosphorylation reaction, the high energy phosphate bond being transferred from the diphosphate to ADP thus forming 3.phospho-glyceric acid and ATP.

    Note: This is the first energy yielding stage, one molecule of ATP being produced for each molecule of 1.3.diphospho-glyceric acid converted to 3.phospho-glyceric acid. It should be noted that in step 4, for each molecule of fructose.1.6.diphosphate, two molecules of 3.phospho-glyceric aldehyde are formed, and since one molecule of fructose.1.6.diphosphate is formed from one molecule of glucose, then for every molecule of glucose entering the system two molecules of ATP are formed in step 6.

    The two molecules of ATP used in the conversion of glucose to fructose.1.6.diphosphate (steps 1 to 3) may be considered to be 'replaced' by the two molecules of ATP formed in step 6, therefore any energy generated in further steps and stored in ATP will be profit to the plant.

(continued on page 49)

TABLE SHOWING REACTIONS AND THE ENZYMES INVOLVED IN ANAEROBIC RESPIRATION

Stage 1

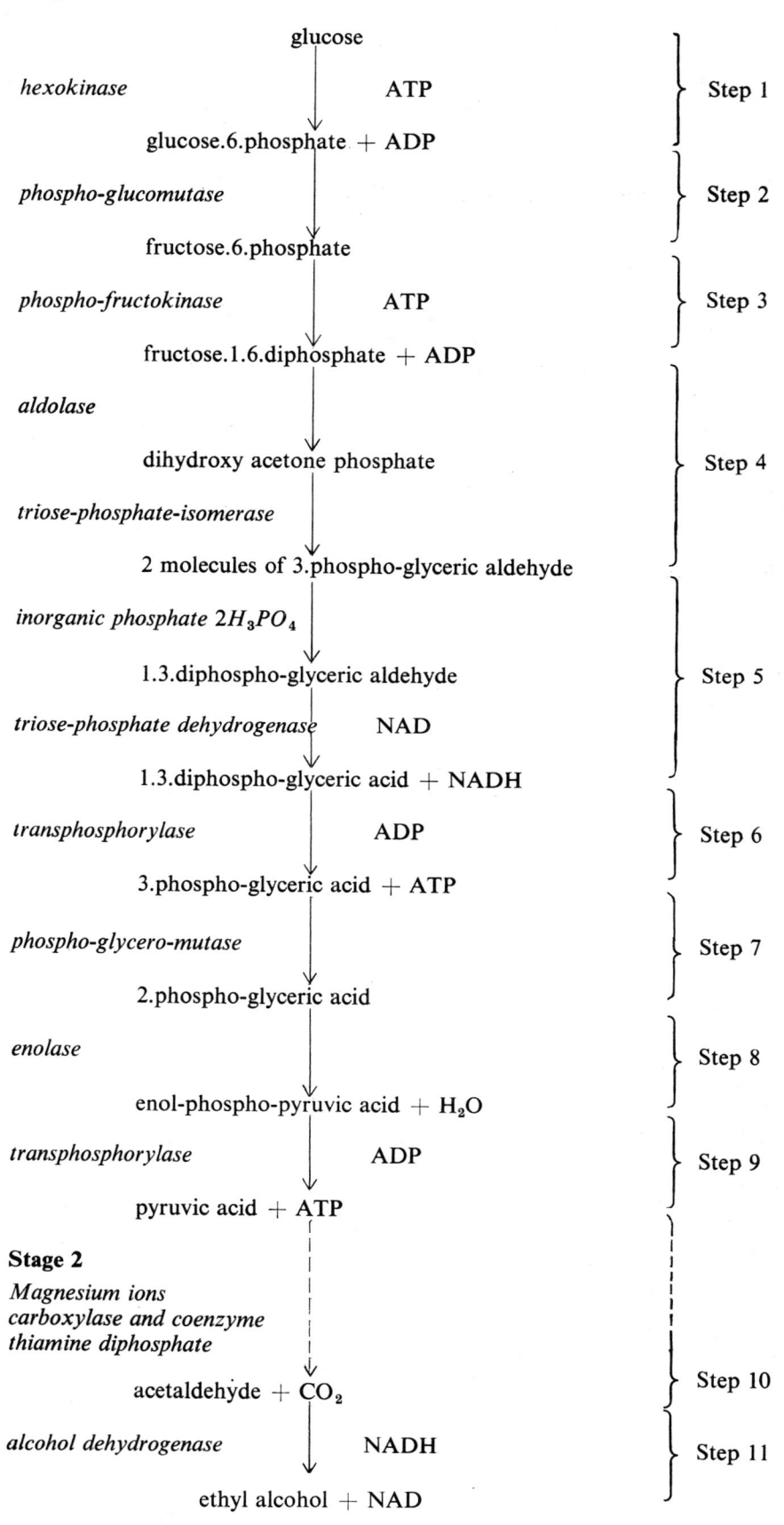

glucose

hexokinase   ATP   Step 1

↓

glucose.6.phosphate + ADP

phospho-glucomutase   Step 2

↓

fructose.6.phosphate

phospho-fructokinase   ATP   Step 3

↓

fructose.1.6.diphosphate + ADP

aldolase

↓

dihydroxy acetone phosphate   Step 4

triose-phosphate-isomerase

↓

2 molecules of 3.phospho-glyceric aldehyde

inorganic phosphate $2H_3PO_4$

↓

1.3.diphospho-glyceric aldehyde   Step 5

triose-phosphate dehydrogenase   NAD

↓

1.3.diphospho-glyceric acid + NADH

transphosphorylase   ADP   Step 6

↓

3.phospho-glyceric acid + ATP

phospho-glycero-mutase   Step 7

↓

2.phospho-glyceric acid

enolase   Step 8

↓

enol-phospho-pyruvic acid + H_2O

transphosphorylase   ADP   Step 9

↓

pyruvic acid + ATP

Stage 2
Magnesium ions
carboxylase and coenzyme
thiamine diphosphate

↓

acetaldehyde + CO_2 Step 10

alcohol dehydrogenase   NADH   Step 11

↓

ethyl alcohol + NAD

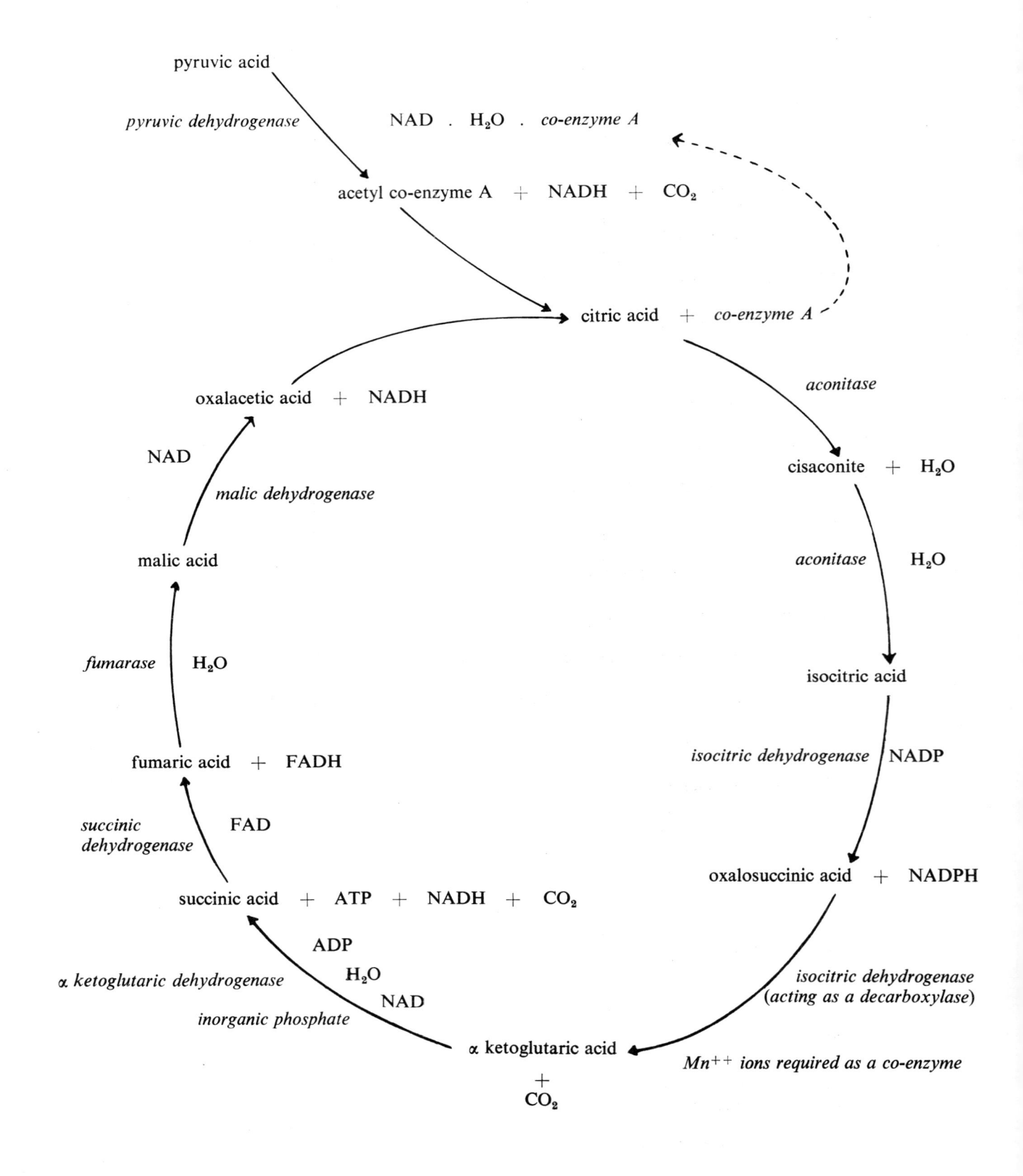

DIAGRAM SHOWING STAGES INVOLVED IN THE KREBS' CITRIC ACID CYCLE

pyruvic acid

pyruvic dehydrogenase $NAD \cdot H_2O \cdot$ *co-enzyme A*

acetyl co-enzyme A + NADH + CO_2

citric acid + *co-enzyme A*

aconitase

cisaconite + H_2O

aconitase H_2O

isocitric acid

isocitric dehydrogenase | NADP

oxalosuccinic acid + NADPH

*isocitric dehydrogenase
(acting as a decarboxylase)*

Mn^{++} *ions required as a co-enzyme*

α ketoglutaric acid
+
CO_2

α *ketoglutaric dehydrogenase*

ADP
H_2O
NAD
inorganic phosphate

succinic acid + ATP + NADH + CO_2

succinic dehydrogenase   FAD

fumaric acid + FADH

fumarase | H_2O

malic acid

NAD

malic dehydrogenase

oxalacetic acid + NADH

(*continued from page* 46)

Step 7. A trans-phosphorylation reaction when 3.phospho-glyceric acid is converted to 2.phospho-glyceric acid.

Step 8. The dehydration of phospho-glyceric acid to form enol-phospho-pyruvic acid and water.

Step 9. A transphosphorylation reaction, the phosphate group being transferred from the enol-phospho-pyruvic acid to ADP, with the formation of pyruvic acid and ATP.

Note: One molecule of ATP is produced for each molecule of 'enol' so, as in stage 6, two molecules of ATP are produced for every molecule of glucose entering the system.

Stage 2.

Step 10. The de-carboxylation of pyruvic acid with the formation of acetaldehyde and carbon dioxide.

Step 11. The reduction of acetaldehyde to ethyl alcohol. The enzyme catalysing this process requires reduced NAD which is produced during step 5. During step 11, the reduced NAD becomes re-oxidised and is then available to act as a hydrogen acceptor again in step 5.

AEROBIC RESPIRATION

This process consists of a series of reactions by means of which glucose is broken down, with the ultimate formation of carbon dioxide and water. The energy liberated during these reactions is stored in high energy bonds in ATP.

The process of aerobic respiration may be divided into three stages which are summarised below.

Stage 1

The glucose molecule is split to form pyruvic acid. During this stage two molecules of ATP are converted to ADP but four molecules of ATP are formed from ADP, therefore a net gain of two ATP.

The reactions in this stage are identical with those occurring in the first stage of the anaerobic process and are referred to as the common path, for details of reactions see stage 1 of the anaerobic process, page 46, also table on page 47.

It should be noted that two molecules of NADH are produced during this stage. In anaerobic respiration these are used in the reduction of acetaldehyde to ethyl alcohol. In aerobic respiration however they are passed on to the third stage where the hydrogen ions or electrons are passed along the respiratory chain.

Stage 2 Krebs' citric acid cycle

This stage is referred to as the **Krebs' Citric Acid Cycle** since it was first postulated by Sir Hans Krebs in 1937. It consists of a complex series of reactions in a cyclic system. The pyruvic acid is converted into acetyl co-enzyme A in the presence of NAD, water, pyruvic dehydrogenase and co-enzyme A. Carbon dioxide and NADH are formed at the same time. The acetyl co-enzyme A combines with oxalacetic acid to form citric acid. The citric acid then enters into the series of reactions referred to as the citric acid cycle by means of which it is converted back to oxalacetic acid and can be used again. The co-enzyme A is also regenerated. Carbon dioxide is formed as a waste product and leaves the system. As the citric acid is broken down, hydrogen ions (or electrons) are removed from the intermediates by enzymes, and, in combination with various hydrogen carriers, leave the system passing to the third stage referred to as the respiratory chain, or hydrogen transport system.

Although most of the acetyl co-enzyme A entering the citric acid cycle is derived from pyruvic acid some is formed from acetic acid derived from fat and amino acid molecules. Fat and protein molecules are, by various enzyme systems not dealt with here, broken down to form acetic acid which after conversion to acetyl co-enzyme A also enters the citric acid cycle.

In all five pairs of hydrogen ions are obtained from one pyruvic acid molecule during its oxidation and degradation in the citric acid cycle, and since each molecule of glucose yields two molecules of pyruvic acid for every molecule of glucose entering the system, ten pairs of hydrogen ions or electrons are passed on to the third stage from the citric acid cycle.

Stage 3

It is during this third and final stage in the respiratory process that most of the energy obtained from the breakdown of glucose is recovered. This stage may be referred to as the energy transfer stage, or as the hydrogen transport system, or as the respiratory chain.

It is composed of a series of three cyclic processes linked by intermediate steps, each part being set in motion

(*continued on page* 51)

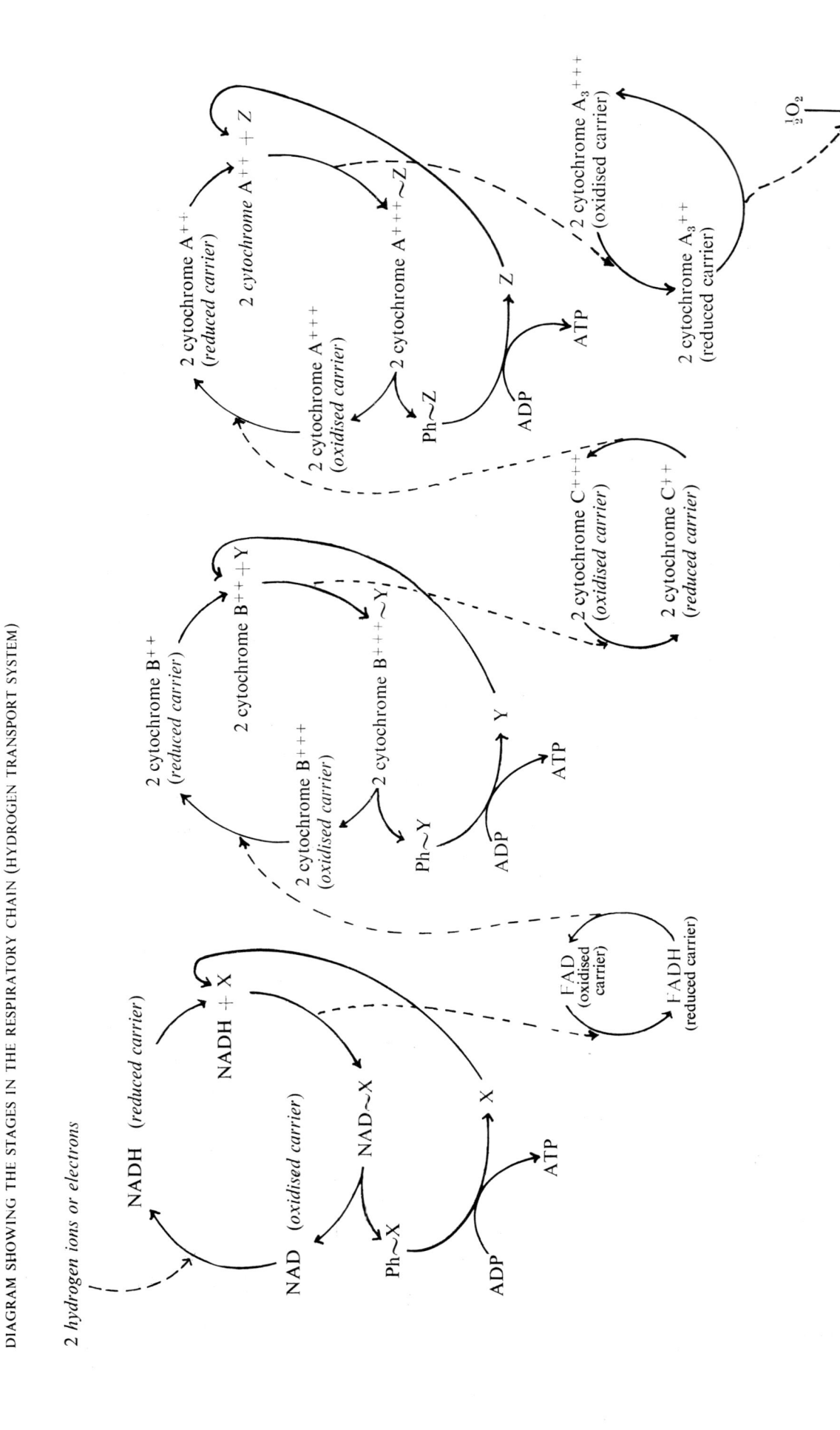

DIAGRAM SHOWING THE STAGES IN THE RESPIRATORY CHAIN (HYDROGEN TRANSPORT SYSTEM)

Note: X, Y and Z are unknown enzymes and $\sim$ indicates the presence of a high energy bond. Dotted lines indicate the links between the cycles as the hydrogen ions or electrons are passed along the chain.

(*continued from page* 49)

by the one preceding. The electrons or hydrogen ions liberated in the citric acid cycle are passed along a chain of electron carrying compounds by a series of oxidation and reduction reactions until they eventually combine with oxygen, the last hydrogen acceptor in the chain, to form water. The high energy electrons are converted, in a stepwise manner into low energy electrons in water. At the same time the energy which is "released" from the electrons is transferred to ADP in combination with phosphate groups to form ATP. In each cycle the acceptor is first reduced by the addition of hydrogen atoms or electrons, and then returned to its oxidised form by oxidation, as the hydrogen is passed on, via the intermediate stage, to the next cycle in the chain.

The exact identity of all the compounds and enzymes involved is as yet uncertain but it is known that each of the acceptors possesses a characteristic active group, capable of accepting electrons from the preceding member of the chain and passing them on to the next, and that each group contains either a metal, such as iron, or a vitamin such as Riboflavin (vitamin B_2).

The exact way in which the hydrogen ions or electrons are transferred is not fully understood, but it is believed that a high energy bond is formed with the enzyme, which then combines with a phosphate group, which is subsequently passed to ADP in the formation of ATP, in which the third phosphate group is attached by a high energy bond. The enzyme is then in combination with the hydrogen carrier, and it is at this stage that the hydrogen ions are passed on to the following intermediate stage, while the cycle is repeated with the entry of two more hydrogen ions and the formation of a high energy bond with the enzyme. The hydrogen ions are passed from the intermediate stage to the following cycle. After the third intermediate stage they combine with oxygen to form water. The passage of each pair of electrons along the chain results in the formation of three molecules of ATP, however one pair of electrons enters the chain at the second cycle and therefore only produces two molecules of ATP.

Energy account

There is a net gain of two molecules of ATP in stage 1, and two pairs of electrons (in NADH) are passed on to the third stage. Ten pairs of electrons are passed on to the third stage from stage two (the citric acid cycle). Thus a total of twelve pairs of electrons pass along the respiratory chain (stage three), one pair however enters at the second cycle producing only two molecules of ATP, while the other eleven produce three molecules of ATP each. Thirtyfive molecules of ATP are therefore produced in the third stage. It will be noted that one molecule of ATP is produced during the second stage. Thus the number of molecules of ATP which are produced from the degradation of one molecule of glucose are as follows:

	two from stage one
	one from stage two
	thirty five from stage three
total =	thirty eight molecules of ATP

Respiratory Quotient

The respiratory quotient (RQ) is the ratio of the carbon dioxide produced to the oxygen consumed in the same time and can be calculated using either the volumes of the gases, or the number of molecules of the gases. By working out the respiratory quotient, it is possible to determine the type of respiration occurring in the tissue or organ, and the percentage of each, if both anaerobic and aerobic are occurring simultaneously, or the type of substrate being used.

RESPIRATORY QUOTIENT FOR AEROBIC RESPIRATION

Overall "equation" for aerobic respiration:—

$$C_6H_{12}O_6 + 6O_2 \rightarrow 6CO_2 + 6H_2O$$

According to the "equation", one volume or one molecule of oxygen is consumed at the same time as one volume or one molecule of carbon dioxide is produced.

$$\text{The respiratory quotient} = \frac{\text{volume (or number of molecules) of carbon dioxide produced}}{\text{volume (or number of molecules) of oxygen consumed}} = \frac{1}{1} = 1$$

Thus the respiratory quotient for the aerobic respiration when glucose is the substrate is 1.

The respiratory quotient varies depending on the substrate. The respiratory quotient for sugars is 1, as mentioned above, for fats it is about 0·7 and for amino acids about 0·8. Variations in the respiratory quotient will also occur when more than one substrate is being used at one time and when both types of respiration occur simultaneously.

THE RESPIRATORY QUOTIENT FOR ANAEROBIC RESPIRATION

Overall "equation" for anaerobic respiration is:—

$$C_6H_{12}O_6 \rightarrow 2C_2H_5OH + 2CO_2$$

According to the equation no oxygen is used but two volumes or molecules of carbon dioxide are produced.

$$\text{the respiratory quotient} = \frac{2}{0} = \text{infinity}$$

Within the range 0°C to 40°C, the increase in the rate of respiration with increase in temperature is approximately exponential, close enough for acceptance within the limits of biological measurements. Taking it as exponential, the ratio of the rates at any fixed temperature interval within the limits, would be constant. In biological measurements a convenient interval is 10°C, from which the symbol Q_{10} is derived referring to the ratio of the rates for this temperature interval.

$$Q_{10} = \frac{\text{the rate at } (t + 10)°C}{\text{the rate at } t°C}$$

Factors affecting the rate of respiration

The overall rate of respiration obviously depends on the size of the plant, the number of actively respiring cells which it contains and its age, but there are also other factors which can cause variations in the rate of respiration.

CHANGE OF TEMPERATURE

Q_{10} (see respiratory quotient above) is approximately 2 between 0°C and 40°C, above this temperature the respiratory enzymes are gradually inactivated and the rate is obviously decreased.

THE PRESENCE OF ADEQUATE WATER SUPPPLIES

It is believed that the water content of the cells probably affects the respiratory activity of the protoplasm as well as affecting the amount of soluble respiratory substrate. The rate of respiration decreases during the drying out of seeds and is increased when they once again take up water.

CONCENTRATION OF SUBSTRATE

The rate of respiration will obviously decrease if the concentration of glucose, the respiratory substrate, is reduced below the optimum concentration and will increase if the concentration then increases, and will continue to increase until optimum concentration is reached. In the case of aerobic respiration a plant can use proteins and fats as respiratory substrates if the glucose concentration is low.

OXYGEN CONCENTRATION

Low concentration of oxygen will affect aerobic respiration since there will be nothing to act as the final acceptor of hydrogen ions at the end of the hydrogen transport system. In such conditions respiration can continue anaerobically temporarily, but an oxygen debt is built up.

RESPIRATION EXPERIMENTS

Experiment 38
To demonstrate anaerobic respiration

APPARATUS REQUIRED
2 conical flasks, each fitted with a rubber bung, one having 1 hole through it, the other having 2 holes through it
2 pieces of bent glass tubing with right angle bends
thistle funnel
lime water
10% glucose solution which has a trace of K_2HPO_4
baker's or brewer's yeast (strains of Saccharomyces cerevisiae)
soda lime
strong coloured solution of iodine in potassium hydroxide
white petroleum jelly

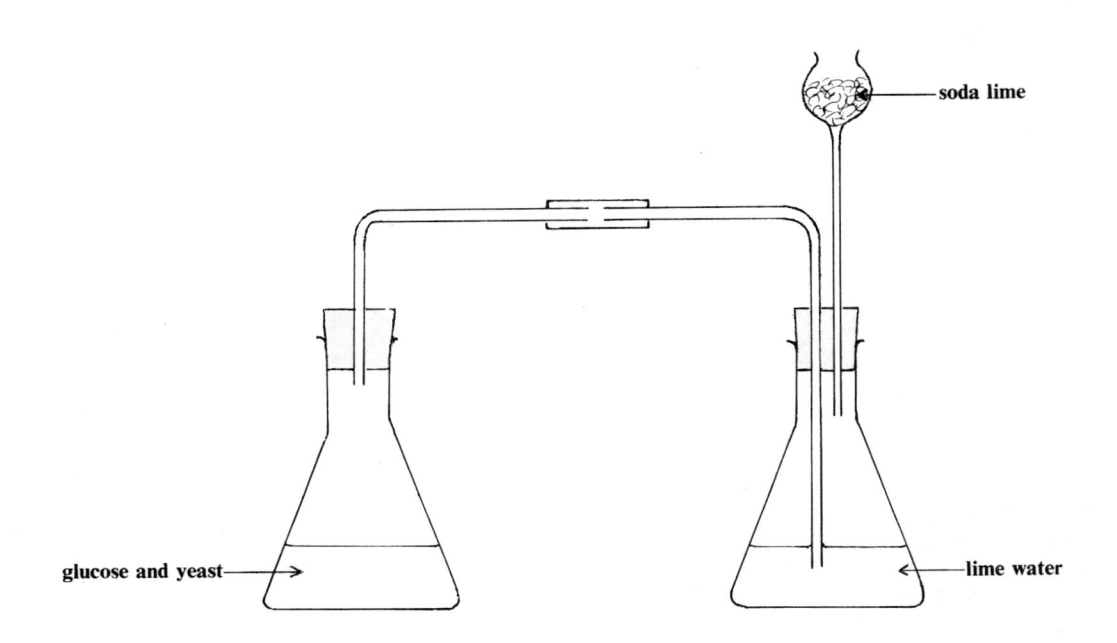

METHOD OF PROCEDURE
Half fill one of the flasks with glucose solution, add to this the fresh yeast. Half fill the other flask with lime water. Insert the stem of the thistle funnel in one of the holes in the two holed bung and insert the long limb of one of the pieces of bent glass tubing in the other hole, then place the bung in the neck of the flask containing the lime water. Make sure that the end of the piece of bent glass tubing is below the surface of the lime water and that the end of the stem of the thistle funnel is above the surface of the lime water. Fill the bulb of the funnel with soda lime. Insert the short limb of the other piece of bent glass tubing into the hole in the other bung and insert the bung in the neck of the flask containing the glucose and yeast, making sure that the end of the piece of glass tubing is well above the level of the liquid in the flask. Connect the open ends of the pieces of bent glass tubing using a piece of narrow rubber tubing making sure that the joints are air tight, if necessary cover with a film of white petroleum jelly. Leave for twenty-four hours then note any change in the lime water, and test the liquid in the other flask for the presence of alcohol using the iodoform test.

IODOFORM TEST
Place about 10 ml of the solution to be tested in a test tube and heat it gently, then add gradually, drop by drop, a strong coloured solution of iodine in potassium hydroxide. The presence of alcohol in the original solution is indicated by the formation of a yellow crystalline precipitate of iodoform accompanied by its characteristic smell.

Note: The soda lime in the funnel prevents any carbon dioxide, from outside the apparatus, entering the flask containing the lime water.

Experiment 39

To demonstrate that anaerobic respiration can occur in tissues normally respiring aerobically

APPARATUS REQUIRED
pea seeds which have been soaked in water for the previous 24 hours
mercury
potassium hydroxide solution
glass dish with vertical sides
test tube
clamp and stand
piece of bent glass tubing
dark paper or cloth
forceps

Diagram of apparatus at the beginning of the experiment, dark cloth not shown

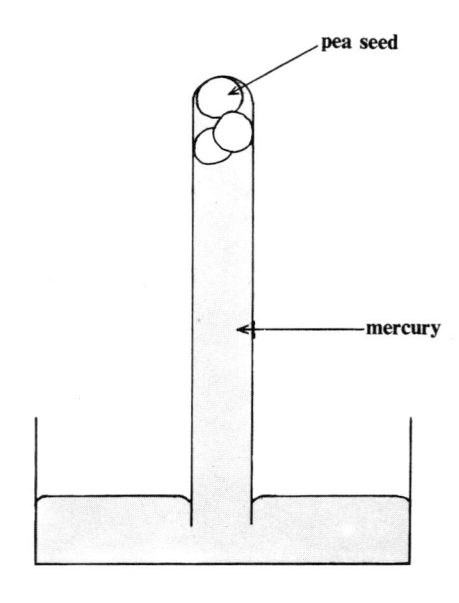

METHOD OF PROCEDURE
Half fill the dish with mercury, then fill the test tube with mercury. Seal the open end of the test tube with your thumb and invert the test tube in the dish of mercury, supporting the test tube by a clamp and stand. Using a pair of forceps, insert one of the peas in the mouth of the test tube, it will immediately rise to the top of the test tube. Repeat this until about four peas have collected in the top of the test tube. Cover the apparatus with a piece of dark cloth or dark paper to prevent the peas photosynthesising. Leave the apparatus for a few days after which note if there has been any change in the level of the mercury in the test tube, and if so determine the nature of the gas in the space formerly occupied by the mercury, by introducing some potassium hydroxide solution into the test tube using a piece of bent glass tubing.

Note: Carbon dioxide is very soluble in potassium hydroxide solution. If the gas at the top of the test tube is carbon dioxide, when the potassium hydroxide is introduced then the carbon dioxide will be absorbed and the mercury will rise to fill the test tube again.

Experiment 40

To demonstrate the production of carbon dioxide as a result of aerobic respiration

APPARATUS REQUIRED
barley grains (or pea seeds) which have been soaked in water overnight
15% potassium hydroxide solution
water
2 glass beakers
2 retorts
2 clamps and stands
sticky tape to mark level of liquids

METHOD OF PROCEDURE
Place some of the barley grains (or pea seeds) in each of the retorts (see following note). Fill one of the beakers with water and the other with the potassium hydroxide solution. Invert the retorts, one over each beaker and arrange them so that the open ends are below the level of the liquids in the beakers. Carbon dioxide is very soluble in potassium hydroxide, and the carbon dioxide present in the air in the retort over the potassium hydroxide is

Diagram showing retorts at the beginning of the experiment

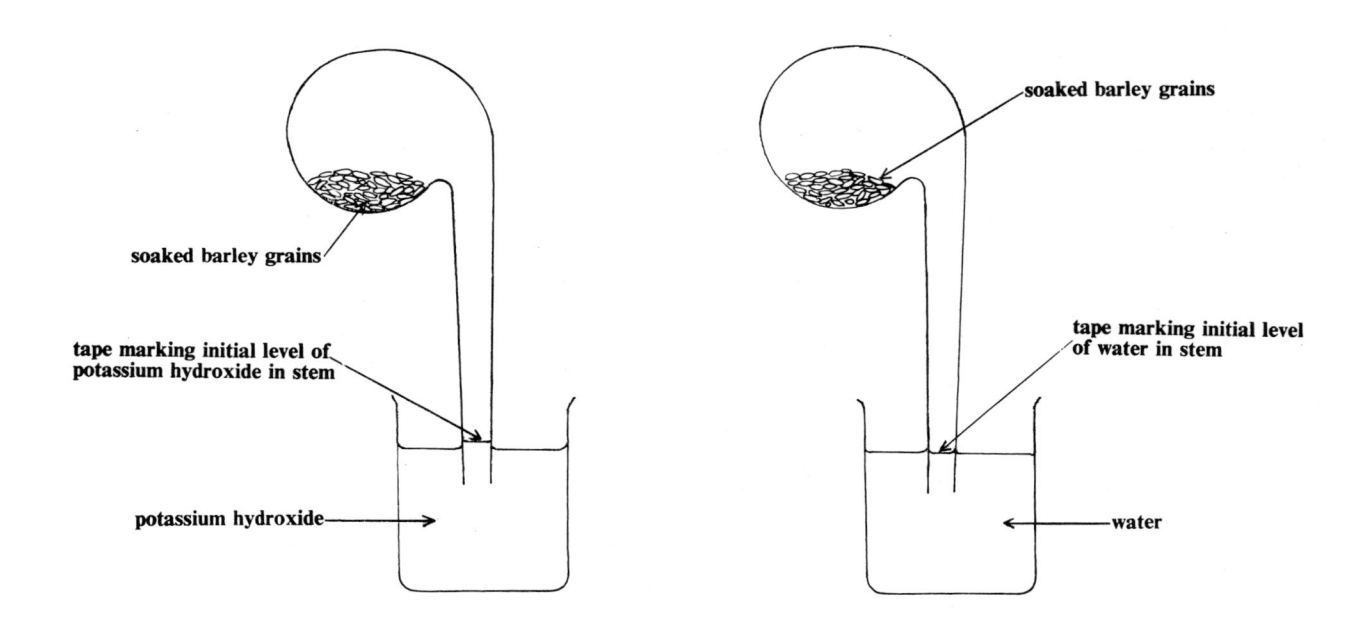

immediately absorbed, so that the liquid rises fractionally up the stem of the retort. Since none of the gases present in air are particularly soluble in water the water does not rise up the stem of the other retort. Mark the level of the liquids in the stems of the retorts with pieces of sticky tape. Leave both pieces of apparatus for several days then note any further change in the levels of the liquids in the stems of the retorts and determine whether respiration has occurred.

Note: If green pea seeds are used then the retorts must be covered with dark cloths to prevent the seeds photosynthesising. If carbon dioxide is produced then it will be absorbed by the potassium hydroxide solution which will rise further up the stem of the retort, but it will not be absorbed by the water which will result in an increase in the volume of air in that retort.

This experiment may be modified to show that respiration is unaffected by the presence or absence of light; in which case 3 retorts should be used, 2 of which are placed over potassium hydroxide 1 of these being left in the light and 1 being placed in the dark. It is essential to use barley grains in this experiment to prevent interference by photosynthesis.

Experiment 41

To demonstrate the production of heat during respiration

APPARATUS REQUIRED
2 vacuum flasks each fitted with a cork which has a single hole through it
pea seeds which have been soaked in water for the previous 24 hours
2 thermometers
beaker of boiling water on a wire gauze and tripod over a bunsen burner

METHOD OF PROCEDURE
Take half of the soaked peas and boil them in water for a few minutes to kill them, after which they should be cooled and then placed in one of the vacuum flasks. Take the rest of the peas, which will still be alive, and place them in the other vacuum flask. Insert the thermometers in the holes in the corks and place a cork, with thermometer, in each flask. Note the initial temperature in each flask. Leave for 12 hours or more then note any change in temperature in either flask.

Note: In a plant energy from respiration is used in growth and in the metabolic processes within the plant but some is lost as heat especially when food reserves are being used rapidly as in a germinating seed.

Experiment 42

To determine the respiratory quotient for barley grains

APPARATUS REQUIRED

conical flask with a side arm (buchner flask)
rubber bung, with a single hole through it, to fit the flask
thermometer
thin walled phials containing caustic soda
manometer with scale
short piece of thin walled rubber tubing
moist barley grains

METHOD OF PROCEDURE

Place the moist barley grains in the buchner flask. Insert the thermometer in the hole in the rubber bung. Tie the phials of caustic soda together and suspend them inside the flask by means of a piece of cotton and hold them in place by inserting the bung, with the thermometer, in the neck of the flask. Connect the side limb of the buchner flask to the manometer by means of the piece of thin walled rubber tubing. Note the level on the manometer, then leave the apparatus for a few days. Note any change in the level on the manometer then shake the flask gently so that the phials of caustic soda break against the wall. Note any further change in the level on the manometer and from your results calculate the respiratory quotient for the barley grains.

Note: The barley grains would normally respire aerobically but in this experiment they are in a confined space so that a certain amount of anaerobic respiration will also take place. The respiratory quotient for aerobic respiration is 1 and the respiratory quotient for anaerobic respiration is infinity, the recorded respiratory quotient in this experiment will be between these two.

SPECIMEN CALCULATION

initial manometer reading   760 mm
second manometer reading   770 mm
third manometer reading   750 mm

initial volume of gases in flask due to nitrogen + oxygen = V = a

volume of gases in flask after two days due to nitrogen + less oxygen + carbon dioxide

$$= V \times \frac{770}{760} = b$$

volume of gases in flask after removal of carbon dioxide due to nitrogen + less oxygen

$$= V \times \frac{750}{760} = c$$

$$\text{Respiratory quotient} = \frac{\text{volume of carbon dioxide produced}}{\text{volume of oxygen used}}$$

$$= \frac{b - c}{a - c}$$

$$= \frac{\dfrac{770\,V}{760} - \dfrac{750\,V}{760}}{\dfrac{760\,V}{760} - \dfrac{750\,V}{760}}$$

$$= \frac{770 - 750}{760 - 750}$$

$$= \frac{20}{10} = 2$$

Experiment 43

To demonstrate the effect of heat on the rate of respiration

APPARATUS REQUIRED
brom-cresol indicator (brom-cresol purple)
pea seeds which have been soaked in water for the previous 24 hours
test tubes
glass tubing
water bath on a tripod, over a bunsen burner
two thermometers

METHOD OF PROCEDURE

Add 25 drops of the brom-cresol indicator to 100 ml of distilled water, then add tapwater drop by drop until the solution turns a pale wine red colour. Place some of this solution in a test tube and pass exhaled air through it using a piece of bent glass tubing. The solution will change colour from wine red to yellow. When the solution is yellow, cork the test tube and keep it as a standard solution. Remove the testa from one of the soaked peas and place the pea in a test tube, add 10 ml of the original wine red indicator solution, note the temperature of the indicator solution ($t°C$) and the time. Leave the test tube, shaking it occasionally, until the indicator solution in it has changed to the same yellow colour as the standard solution. Note the time taken for this change to take place.

Heat the water in the water bath to a temperature 10 degrees above $t°C$. Place 10 ml of the original wine red indicator solution in a test tube and then place this in the water bath and heat it until it is at the temperature $(t + 10)°C$. At the same time remove the testa from a second pea, place the pea in 10 ml of water in a test tube and heat this in the water bath to $(t + 10)°C$. When both indicator and pea are at the correct temperature transfer the pea to the indicator solution then note the time taken for the pea to attain the standard yellow colour. Compare the time taken at the different temperatures and determine Q_{10}.

$$Q_{10} = \frac{\text{rate at } (t + 10)°C}{\text{rate at } t°C}$$

since the rate is equal to the inverse of the time taken

$$\text{then } Q_{10} = \frac{\text{time at } t°C}{\text{time at } (t + 10)°C}$$

BIOCHEMISTRY

This section is for convenience divided into two sub-sections, the first of which deals with the chemistry of the more important plant products, while the second describes the general properties of the enzymes involved in the metabolism of the plant.

THE CHEMISTRY OF PLANT PRODUCTS

A very large number of different compounds occur in plants, and, strictly speaking, since most of them are essential for the well being of the plant, they cannot be differentiated into important and non-important compounds. However, by virtue of the fact that they are either present in large amounts or are involved in reactions which interest scientists, certain types of compounds may be considered to be "more important" than others. Included in this category are the carbohydrates, the proteins, the lipids, and the nucleic acids.

THE CARBOHYDRATES

In the same way that most of the bulk of an animal consists of proteins, so most of the bulk of a plant consists of carbohydrates. These have a wide variety of roles within the plant, and, as their name implies, they consist of the elements carbon, hydrogen, and oxygen.

The carbohydrates are generally classified into three main groups:— the **monosaccharides**, the **oligosaccharides**, and the **polysaccharides.**

Monosaccharides

The monosaccharides are simple carbohydrates and cannot be hydrolysed into smaller units. They have the general formula $C_n H_{2n} O_n$. The actual number of carbon atoms present varies from three to eight, and it is on the basis of this that the monosaccharides are subdivided. A sugar having only three carbon atoms is described as a **triose**, one having four carbon atoms as a **tetrose**, five as a **pentose**, six as a **hexose**, seven as a **septose**, and eight as an **octose**.

Although between sixty and seventy different monosaccharides are known, only a small number actually occur in plants and most of these are present in very small amounts. The only monosaccharides which occur in amounts easily detectable and which are universally present in the cytoplasm of plant cells are the hexose sugars, **D-glucose** and **D-fructose**. Although these have the same general formula, $C_6 H_{12} O_6$, they differ in the arrangement of the hydrogen and oxygen atoms, glucose having an aldehyde group at carbon atom number one (C1), and fructose a ketone group at carbon atom number two (C2).

		carbon atom number	D-glucose	D-fructose
aldehyde group	C—H, =O	1	CH(=O)	CH$_2$OH
		2	HCOH	C=O
		3	HOCH	HOCH
ketone group	C=O	4	HCOH	HCOH
		5	HCOH	HCOH
		6	CH$_2$OH	CH$_2$OH

The aldehyde group of glucose and the ketone group of fructose can be readily oxidised by certain reagents and because these reagents themselves become reduced in the process, both glucose and fructose are said to be "reducing sugars". The two sugars can however be distinguished by specific tests for the aldose and ketose groups.

Although glucose and fructose have been shown on the previous page with their carbon atoms arranged in a straight chain, the evidence is that most of the time, the sugars occur in a ring or cyclic form with the type of ring depending on the sugar involved.

D-glucose exists as a **pyranose** or six-membered ring in which an atom of oxygen is used to join carbon 1 with carbon 5. With the formation of the ring a new hydroxyl (OH) group is formed at carbon 1 and the orientation of this group relative to rest of the molecule decides whether α or β-D-glucopyranose results.

β-D-glucopyranose   α-D-glucopyranose

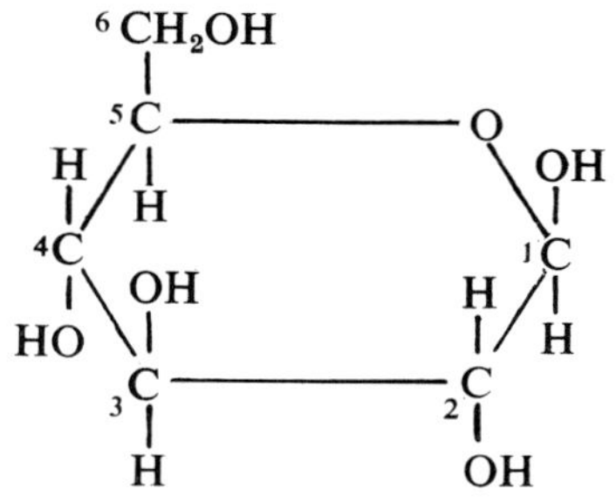

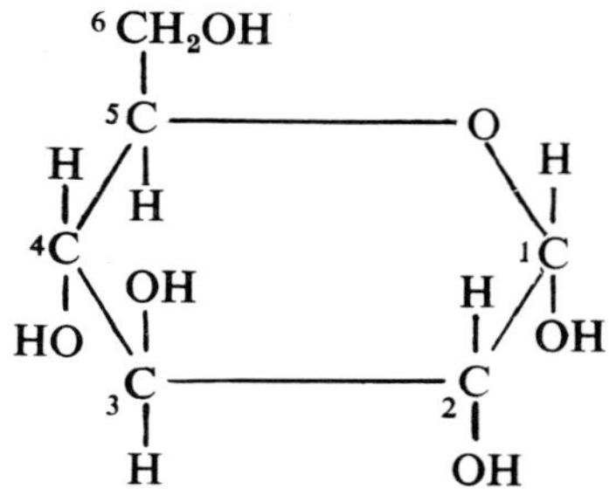

These two forms of glucose have widely differing properties when present in polysaccharides, as we shall see later. **D-fructose** can also occur as a six membered ring, fructopyranose, but more often occurs with the molecule arranged as a furanose or five membered ring, the oxygen bridge in this case linking carbon atoms 2 and 5.

D-fructofuranose

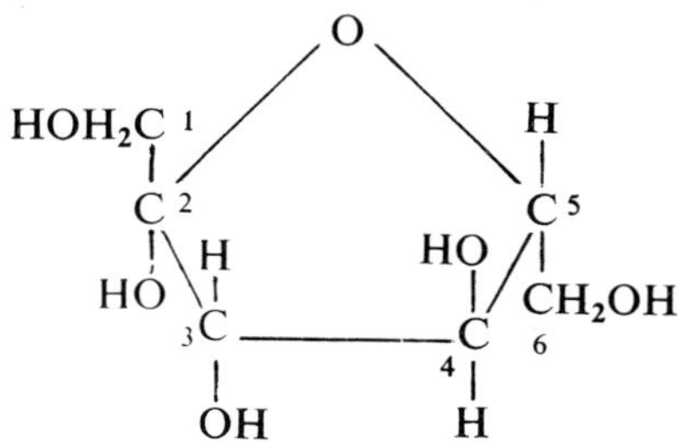

Oligosaccharides

Oligosaccharides are carbohydrates in which two, three, or four monosaccharides are combined together. Such oligosaccharides are described as **disaccharides, trisaccharides** and **tetrasaccharides.**

The **disaccharides** are formed by the condensation of two hexose sugars and have the general formula $C_{12}H_{22}O_{11}$, the link between the monosaccharides having been formed by the elimination of the elements of one molecule of water.

$$2\ C_6H_{12}O_6 - H_2O = C_{12}H_{22}O_{11}$$

E

The most important disaccharide found in higher plants is **sucrose**, which is not only the direct product of photosynthesis, but is also the chief translocatory compound and in some plants such as sugar beet, even serves as a storage carbohydrate.

It has been shown by hydrolysis that sucrose consists of one molecule of glucose (in the α-D-pyranose form) combined with one molecule of fructose (in the furanose form). Although both glucose and fructose, by virtue of their respective aldehyde and keto groups are reducing sugars, sucrose is not, which indicates that the linkage between the two sugars involves these groups.

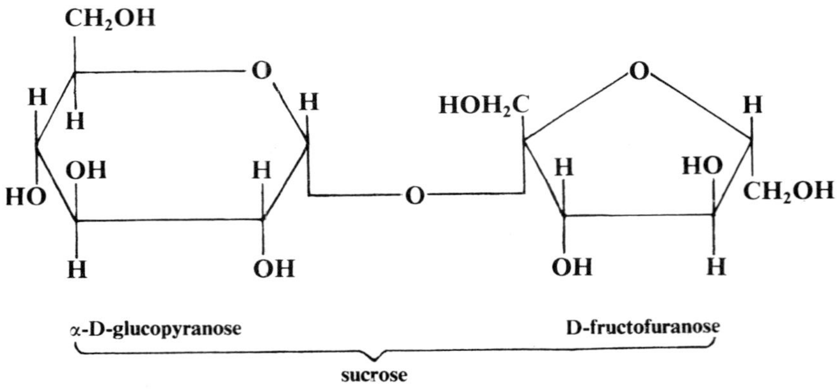

$$\underbrace{\text{α-D-glucopyranose} \qquad \text{D-fructofuranose}}_{\text{sucrose}}$$

Of the other oligosaccharides only the trisaccharide **raffinose** is of widespread occurrence, and even this occurs only in small amounts.

Polysaccharides

These compounds result from the condensation of large numbers of monosaccharides and have the general formula $(C_6H_{10}O_5)n$ where n is the number of monosaccharide units involved.

The polysaccharides occurring in plants fall into two categories namely, the **food storage** compounds such as starch and inulin, which play an important part in the energy economy of the plant, and the **structural** compounds of which cellulose is the best known example.

Starch is the principal reserve carbohydrate of higher plants, large amounts being stored as solid grains in seeds, tubers and roots, where it serves as a source of food for newly developing tissues. Starch also accumulates in leaves during the day as a result of photosynthesis. If the starch grains from various types of plants are examined it will be seen that each has a characteristically shaped grain. It will also be seen that the grains appear "layered". This is due to the depositing of dense starch during the day and less dense starch during the night. If the plants were to be grown under conditions of constant light, the grains would not be layered.

A great deal of work has been carried out on the structure of starch and it is now known to consist of two components, **amylose** and **amylopectin**.

Amylose has a very high molecular weight and consists of a long, unbranched, chain of α-D-glucopyranose molecules with the molecules linked in the α1:4 positions as shown below.

Portion of amylose chain

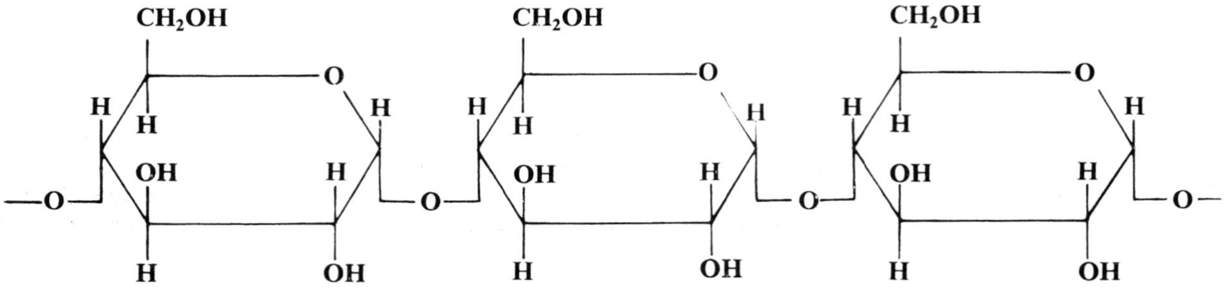

The actual number of glucopyranose units in a single molecule of amylose varies from plant to plant but generally lies between 300 and 1,000. The amylose chain is believed to be wound in a helix, each turn of the helix consisting of six glucose molecules.

Amylopectin has an even higher molecular weight than amylose. Like amylose, it is made up of α-D-glucopyranose units but in this case the molecule is much branched. The branches occur every eight glucose units of the main chain and are some eighteen glucose units long. The branches may be attached to other chains so that a much ramified structure results. The α-glucopyranose units of the branches are linked together by α 1:4 bonds as in the amylose chain and where the branches are attached to the chains, the carbon 1 of the terminal glucose residue of the branch is attached to the number six carbon atom of a glucose unit of the main chain.

Diagram showing link between the glucose unit of the branch and that of the main chain, in amylopectin

The difference in the arrangement of the glucose units in amylose and amylopectin is reflected in their physical and chemical properties. Amylose is more soluble in water than is amylopectin and the two can be separated by standing starch in water for a prolonged period. The amylose will eventually pass into solution while the amylopectin remains as residue.

The two compounds also differ in their reaction to iodine. Amylose reacts to produce the typical blue colour of the starch: iodine reaction, while amylopectin produces a red or violet colour.

While starch is the most widely occurring plant storage product, some higher plants such as Dahlia, Dandelion, and other members of the compositae, produce instead the reserve polysaccharide **inulin**. Inulin resembles amylose in that it consists of a chain of repeated units, but differs in that the repeating units are fructofuranose molecules. It has been estimated that there are some twenty-eight fructofuranose units in every inulin chain and that these units are linked through carbon 1 of one unit and carbon 2 of the succeeding unit.

The chief structural polysaccharide of plants is **cellulose** which makes up the larger part of the cell walls. Cellulose in fact occurs in nature in greater amounts than any other organic compound. As with other polysaccharides, the cellulose molecule consists of a large number of monosaccharide units linked to form a long unbranched chain. The monosaccharide involved is glucose which occurs in the β-D-**glucopyranose** form. Each monosaccharide unit is joined to the next by a β 1:4 linkage so that the only difference between the structure of cellulose and amylose is that the repetitive units of the former are β-D-glucopyranose residues, while in the latter they are α-D-glucopyranose residues. Although these units differ only in the orientation of a single hydroxyl group, the polysaccharides formed by their respective polymerisations have widely differing properties.

Diagram representing a portion of a cellulose molecule

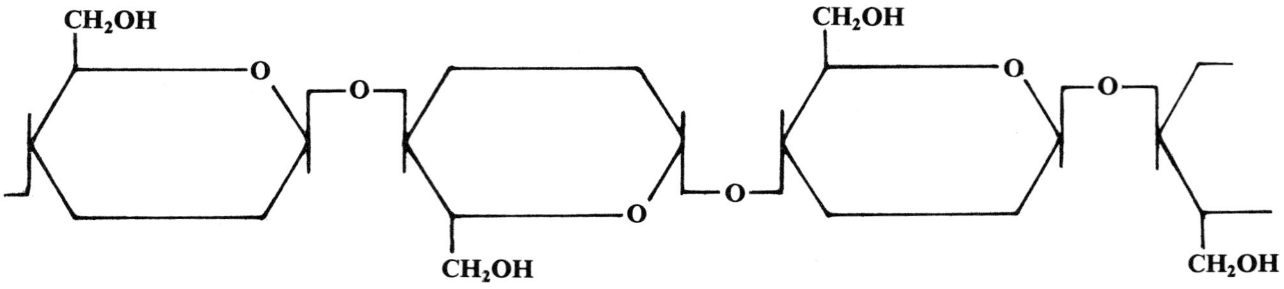

Cellulose occurs in a pure form in the walls of the fibre cells of such plants as cotton and flax, which in fact serve as commercial sources of the compound. Investigations into the structure of the cellulose of these plants has shown that the chains of β-D-glucopyranose residues occur in various lengths, the number of units varying from 1,400 to 10,000. Cellulose is therefore a very high molecular weight polysaccharide.

In the physical structure of cellulose approximately 100 chains of glucopyranose residues are combined to form crystalline bundles or **micelles.** Since each micelle is approximately 120 residues long, a single cellulose molecule participates in the formation of several micelles. Between two micelles the glucose chains remain separate thus forming a non-crystalline region. The micelles are themselves combined to form strands or **microfibrils** and it is the microfibrils which are visible to the electron microscope and which interweave to form the cell wall.

The walls of most plant cells contain, in addition to cellulose, other polysaccharides which occupy the spaces between the micelles and also between the microfibrils. These polysaccharides fall into two categories, the **hemicelluloses** and the **pectic** compounds.

The **hemicelluloses** are a very heterogeneous group of substances, their structure varying according to the source. They serve a dual role in the metabolism of the plant since although they have a chiefly structural function, they may also serve as nutrient reserves in seeds where they are hydrolysed on germination. Like cellulose the hemicelluloses consist of chains of linked monosaccharide units, but the chains are generally shorter than in cellulose and in some cases may be branched. The monosaccharides occurring most frequently are the hexose sugars **D-galactose** and **D-mannose** and the pentose sugars **D-xylose** and **L-arabinose.** Usually only two of these four sugars occur in any one hemicellulose chain.

α-D-galactopyranose

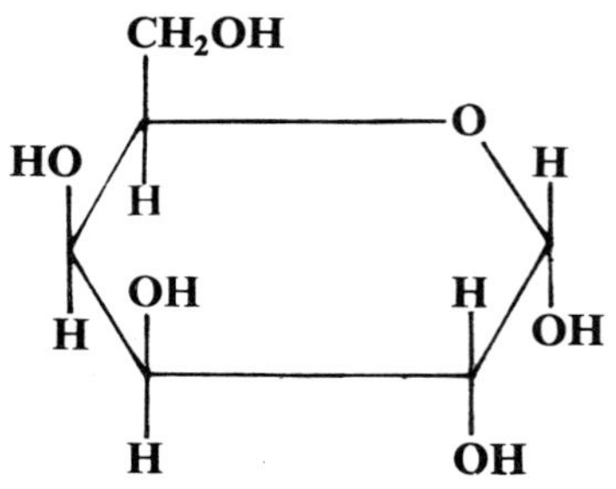

α-D-mannopyranose

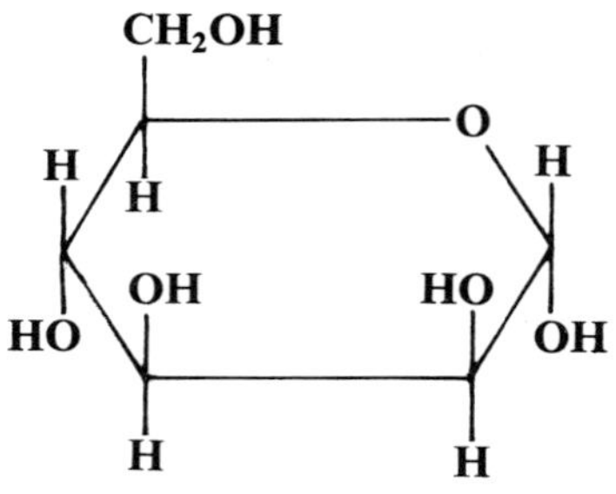

α-D-xylopyranose   α-L-arabinopyranose

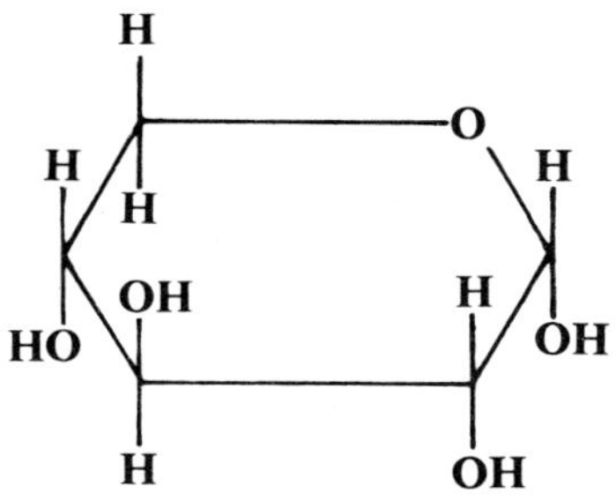

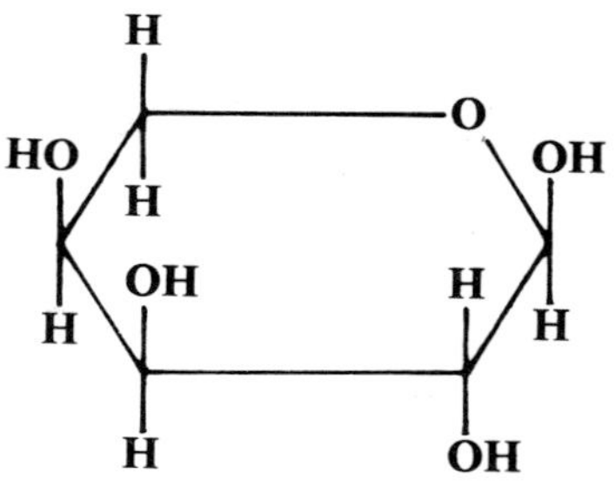

The **pectic** compounds also consist of short branched chains but are distinguished by the fact that the repeating units are **sugar acid residues.** The structure of many of the pectic compounds is still a matter for investigation since they form a very diverse group, but the best known member, **pectic acid** consists of polygalacturonic acid chains.

Pectic acid

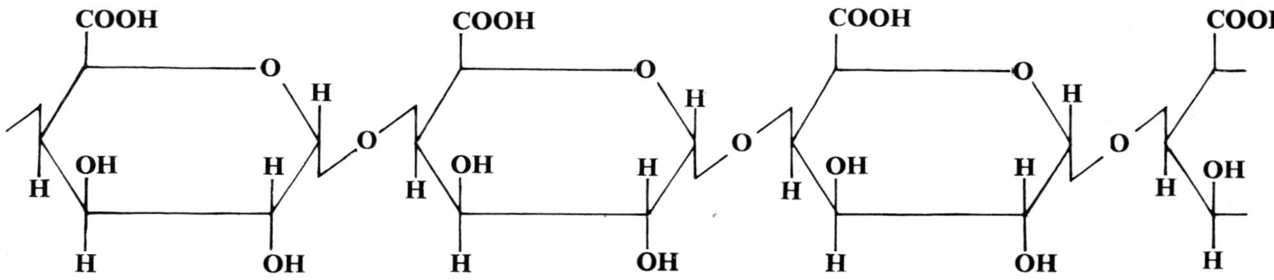

PROTEINS

Many of the biological, chemical and physical properties of living material are the result of the presence of proteins. In general the proteins of the cell fall into two broad groups, the **enzymatic proteins** and the **structural proteins.**

The **enzymatic proteins** are responsible for catalysing the chemical reactions which occur in cells. These reactions would proceed only at a low rate in their absence (see page 71.).

The **structural proteins,** together with the lipids, form the membranes of the cell which not only restrict the diffusion of ions and molecules but also provide a "skeleton" on which enzymes can be arranged in specific orders.

Both the structural and the enzymatic proteins have the same general structure. They are complex molecules with extremely high molecular weights which range from 10,000 to over 1,000,000. Like polysaccharides they are made up of large numbers of **sub-units** polymerised to form chains. The sub-units are **amino-acid** molecules of which some 20 different kinds are known to occur in proteins. The simplest amino-acid is **glycine** which has the formula:—

Glycine $CH_2\ COOH$
              $|$
              NH_2

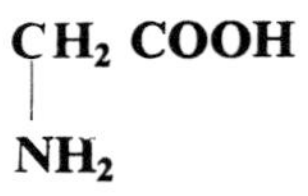

The carbon atom next to the carboxyl group (COOH) is referred to as the α carbon atom and since the NH_2 group is attached to this atom, glycine is an **α-amino acid.**

The amino acids found in proteins are all α-amino acids, and, as such, are structurally similar to glycine, having the general formula:—

α-amino acid $R-\overset{\alpha}{CH}-COOH$ where R = a specific side group.
                              $|$
                             NH_2

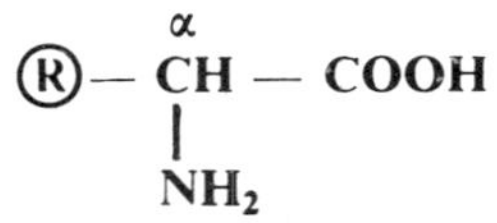

The side group may be a simple methyl group (CH_3) as in **alanine**, or a large group such as the indole group of **tryptophan**.

L-alanine

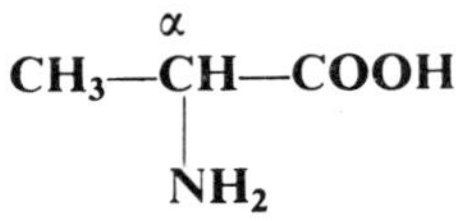

$CH_3-\overset{\alpha}{CH}-COOH$
         $|$
        NH_2

Tryptophan

(indole ring)$-CH_2-\overset{\alpha}{CH}-COOH$
                            $|$
                           NH_2

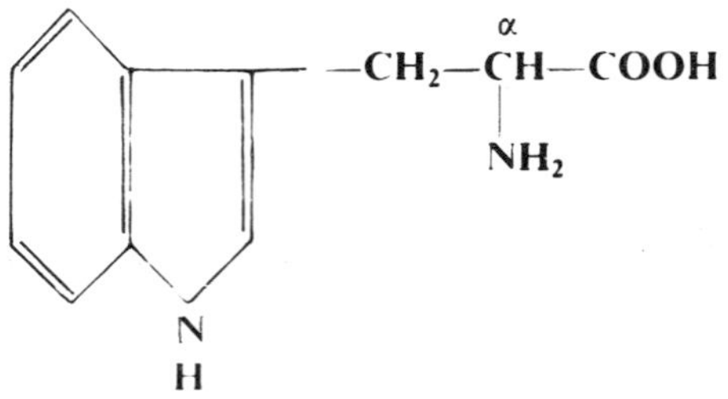

Because they have both an acidic group (COOH) and a basic group (NH_2), amino acids can behave as acids or bases, that is they are amphoteric. In acid solutions they exist as cations $^+H_3N\ CH\ (R)\ COOH$ and in alkaline solutions as anions $H_2N\ CH\ (R)\ COO^-$. In neutral solutions they are present as dipolar ions or Zwitter ions $^+H_3N\ CH\ (R)\ COO^-$.

The amino acids are linked together through the carboxyl group of one molecule and the amino group of the next to form chains or **peptides**. The linkage is known as a **peptide bond**, and is formed by the loss of a molecule of water.

peptide bond $HN-\overset{\overset{O}{\|}}{C}$

amino acid $H_2N-\overset{R_1}{\underset{}{CH}}-CO\ [OH----H]\ HN-\overset{}{\underset{R_2}{CH}}-COOH$ amino acid

condensation reaction ↓

peptide $H_2N-\overset{R_1}{\underset{}{CH}}-\overset{\underset{O}{\|}}{C}-NH-\overset{}{\underset{R_2}{C}}-COOH$ + H_2O

peptide bond

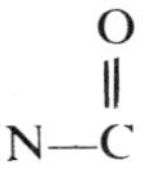

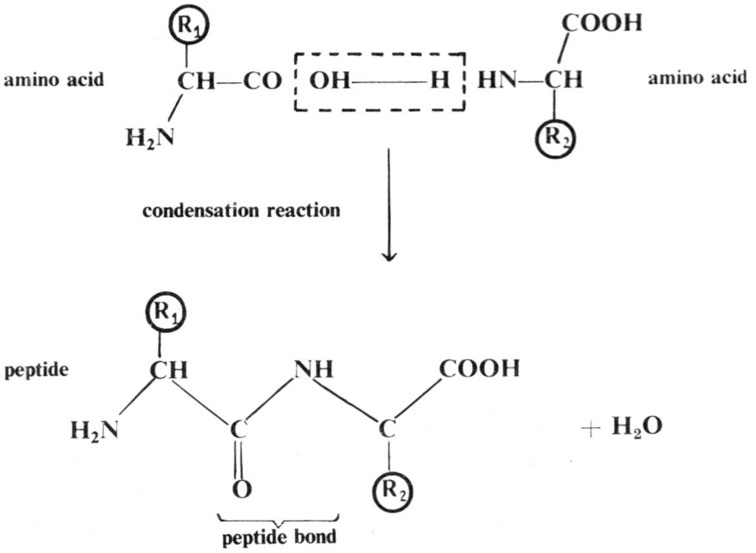

In proteins the chains are very long containing at least 100 amino acids and for this reason proteins are referred to as **polypeptides**. Unlike many polysaccharides, polypeptides never consist solely of a single type of sub-unit, in fact proteins are known which have most of the 20 types of amino acids involved in their structure. Because of the extremely high number of possible combinations of the amino acids, the potential number of different types of proteins is almost incalculable.

The structure of a protein can be examined at various levels:—

a) A study may be made of the amino acids present and their sequence in the polypeptide chain. This is known as the **primary structure** of the protein. The sequence of the amino acids is very important since two proteins may have the same structure over the greater length of their chains but be different at one place and thus have widely separated properties. If, for example, one change is made in the long chain structure of insulin, it becomes physiologically inactive.

b) The arrangement of the amino-acids in space, that is the three dimensional structure also forms a basis for examination. This is the **secondary structure** of the protein and in many polypeptides it takes the form of a right handed helix.

c) A study may also be made of the **tertiary structure** of the protein, which is formed when the long spiral chain is itself coiled and folded to form a compressed structure. Because the R-side chains form linkages between the coils or folds, the protein molecule is fairly rigid. The folding and twisting of the polypeptide chain is not random but is highly organised. The folding results in those amino acids with hydrophilic (water loving) side chains being on the outside where they are in contact with the aqueous phase in which the proteins exist, while the amino acids with hydrophobic side chains are on the inside of the molecule.

All proteins are subject to damage by such agents as heat, strong acids, or bases, various organic compounds and the ions of heavy metals such as lead, silver and mercury. Because the damage changes the nature of the protein and prevents it functioning properly, these agents are said to **denature** protein. Denaturisation by heat is the result of the increased temperature causing interchain linkages to become hydrolysed. The result is that the tertiary structure of the protein becomes changed. Strong acids or bases have a similar action. Heavy metal ions usually cause alteration of the protein structure by becoming attached to part of the protein. In the case of enzymatic proteins, the metal ion usually becomes attached to the "active" part of the molecule and thus prevents the enzyme functioning. A common sign of denaturisation is the precipitation of the protein from a solution. Thermal denaturisation is normally irreversible but denaturisation by metallic ions and salts can often be reversed.

LIPIDS

The lipids are fatty substances, the term covering a wide variety of compounds which are chemically unrelated. They may be divided into three groups as follows:—

a) **simple lipids**—true fats, fatty acids and waxes
b) **conjugated or compound lipids**—phospholipids
c) **derived lipids**—steroids, xanthophylls, carotenoids

The simple lipids are important in the energy economy of the plant and the seeds of many plants. For example, sunflower and linseed have their energy stored, not in the form of polysaccharides, but as fats and oils.

The **phospholipids** and the **steroids** are of structural importance since the membranes of a cell consist of a double layer of lipid material, sandwiched between two protein layers. It is the lipid layer which seems to confer on the membrane many of its permeability properties since, in general, the more soluble a substance is in lipids, then the more easily it passes through the cell membrane. Chemical analysis of cell membranes have shown them to be rich in such phospholipids as **lecithin** and **phosphatidyl serine,** and also the steroid **cholesterol.**

Lecithin

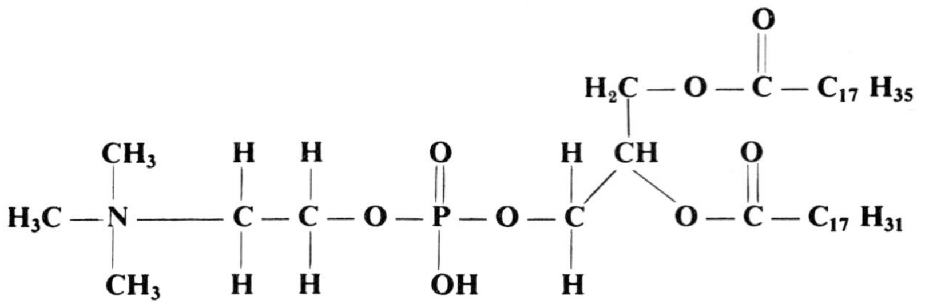

$$H_3C-N\begin{matrix}CH_3\\|\\\\|\\CH_3\end{matrix}\begin{matrix}H&H\\|&|\\-C-C-\\|&|\\H&H\end{matrix}\begin{matrix}O\\||\\O-P-O-\\|\\OH\end{matrix}\begin{matrix}H&CH\\|\\C\\|\\H\end{matrix}\begin{matrix}H_2C-O-\overset{O}{\overset{||}{C}}-C_{17}H_{35}\\\\O-\overset{O}{\overset{||}{C}}-C_{17}H_{31}\end{matrix}\quad\text{fatty acid residues}$$

The value of the phospholipids lies in their dipolar nature. The phosphate group is strongly hydrophilic and the fatty acid residues lipophilic (fat loving). This enables the phospholipid molecule to orientate itself, with the fatty acid residues projecting into the fat, and the phosphate residue projecting into water, thus stabilising the interface between the two phases.

The xanthophylls and carotenes owe their importance to the fact that they act as photosynthetic pigments, absorbing light at frequencies which the chlorophylls cannot absorb, and passing the energy from such light to the chlorophyll molecules.

BIOCHEMICAL TESTS

TESTS FOR CARBOHYDRATES

Before dealing with the sugars actually isolated from the plant, it is advisable that the following tests and reactions should be carried out with pure samples of sugars.

General Tests

1. MOLISCH'S TEST

All carbohydrates, and also compounds containing carbohydrates in a combined form, are degraded to furfural or a derivative of furfural by concentrated sulphuric acid. The furfural, or its derivative, condenses with α-naphthol with the formation of a purple substance.

To 2 ml of sugar solution add 1 drop of 10% α-naphthol in ethanol. Run 1-2 ml of concentrated sulphuric acid down the side of the sloping tube containing the α-naphthol solution, so that it forms a layer at the bottom of the tube. Observe the colour at the interface between the two layers. As a control repeat using water instead of the sugar solution.

Test solutions of glucose, fructose, sucrose and starch.

2. THYMOL TEST

To a small quantity of a sugar solution add a few drops of alcoholic thymol and an excess of concentrated hydrochloric acid. On boiling a carmine colour results. Repeat using water instead of the sugar solution.

Test solutions of glucose, fructose, sucrose and starch.

Tests for Monosaccharides

1. FEHLING'S TEST

To approximately 1 ml of Fehling's solution A, add approximately 1 ml of Fehling's solution B. Add 2 ml of the solution to be tested, mix well and then boil. A red precipitate of cuprous oxide is produced in the presence of a reducing sugar.

Test solutions of glucose, fructose, and xylose.

2. BARFOED'S TEST

To 1-2 ml of Barfoed's reagent add an equal volume of sugar solution and boil for 1 minute. Leave to stand. If reduction has taken place, a red precipitate results.

Test solutions of glucose, fructose and xylose.

3. SELIWANOFF'S TEST

To 2-3 ml of the reagent (0·05% resorcinol in 3N HCl) add 2-3 drops of the sugar solution and heat in a boiling water bath **for 1 minute only.** Compare carefully the colour, and the time of formation, given by the different sugars. Do not heat excessively.

Test solutions of glucose, fructose and xylose.

4. RAPID FURFURAL TEST

To 1 ml of sugar solution, add 1 ml of 10% α-naphthol in ethanol and 6-8 ml of concentrated hydrochloric acid. Mix thoroughly and then boil. Note the time taken for colour development after the mixture starts boiling.

Test solutions of glucose, fructose and xylose.

5. BIAL'S TEST

To 5 ml of orcinol in concentrated hydrochloric acid, add 2-3 drops of sugar solution and then boil. Note the colour formed. Cool and add 2-3 ml of pentanol. The coloured substance is more soluble in pentanol than in the aqueous phase. This test is usually applied for the identification of pentoses.

Test solutions of glucose, fructose and xylose and compare the reactions given by glucose and fructose with that given by xylose.

6.

Boil a mixture of equal volumes of sugar solution and concentrated hydrochloric acid.

Test solutions of glucose and fructose and compare the results.

Tests for Disaccharides

Repeat the tests numbers 1, 2, 3, and 4, for monosaccharides using solutions of sucrose and maltose. Note that sucrose is non-reducing. Explain this.

To a solution of sucrose add some concentrated sulphuric acid and then boil for 2 minutes. Neutralise the mixture with caustic soda using litmus paper as indicator. Repeat tests, numbers 1, 2 and 3 for monosaccharides using the neutralised solution and explain your results.

Tests for Polysaccharides

A. **Starch**

1. PHYSICAL PROPERTIES

Examine some of the starches from rice, wheat, and potato under a microscope. Test the solubility of starch in cold water. Examine the effect of boiling,

2. IODINE TEST

To 2-3 ml of starch solution add 1-2 drops of iodine solution. Compare the colour with a water control. Note the effect of boiling and subsequent cooling.

3. FEHLING'S TEST

Perform the test as on page 66 on the starch solution and note the result. Then to a few millilitres of starch solution add a few drops of dilute hydrochloric acid and then boil the mixture. Cool, neutralise, and re-test with Fehling's solutions. Repeat test using water in place of dilute hydrochloric acid. Explain your results.

B. **Cellulose**

Both filter paper and cotton wool are almost pure cellulose and will serve as a convenient source of experimental material.

1. Dip a little cotton wool into a calcium chloride iodine solution. A rose red colouration is produced which eventually turns violet.

Calcium chloride iodine solution is prepared as follows:— to 10 ml of a saturated solution of calcium chloride add 0·5 g of potassium iodide and 0·1 g of iodine. Warm gently and filter through glass wool.

2. Dip a little cotton wool or filter paper into a solution of iodine in potassium iodide. Place the stained material into an evaporating dish and add a drop or two of concentrated sulphuric acid. A blue colouration results due to the formation of an "amyloid".

3. Place a few small pieces of filter paper into Sweitzer's reagent. Leave the cellulose and reagent for a while and observe the effect.

4. FEHLING'S TEST

Perform this test as on page 66, using small pieces of filter paper or a few threads of cotton wool. Note the result.

Dissolve as much filter paper as possible in 5 ml of concentrated sulphuric acid and when all is in solution, pour into 100 ml of distilled water. Boil the solution for 1 hour in a round bottomed flask, using a sand bath for heating. Neutralise with solid calcium carbonate, add a little water and filter. Repeat the Fehling's test as on page 66. Explain your results.

C. **Inulin**

1. MOLISCH'S TEST—as previously described on page 66.

2. FEHLING'S TEST

Perform the test as on page 66 on a solution of inulin and note the result.

Dissolve some inulin in very dilute hydrochloric acid and heat on a water bath for half an hour. Add more hydrochloric acid from time to time. Neutralise the solution with sodium carbonate and concentrate on a water bath. Repeat the Fehling's test and explain your result.

3. SELIWANOFF'S TEST—as on page 67.

4. Cut sections of dandelion root into 90% alcohol and leave for a short time. Mount the sections in glycerine and examine under a microscope. Crystals of inulin will be present in the cells.

D. **Lignin**

Although this compound is not a carbohydrate it is included here since it is associated with cellulose in the walls of woody cells.

1. Cut sections of a stem or root and immerse in phloroglucinol acidified with hydrochloric acid. The lignin turns red. Note which tissues are affected.

2. Repeat the above experiment using aniline hydrochloride—lignin turns bright yellow with this reagent.

TESTS FOR AMINO ACIDS

General Reactions

1. PHYSICAL PROPERTIES

Test the solubility of glycine and glutamic acid in water, dilute alkali, dilute acid, ethanol and ether.

2. REACTIONS WITH NITROUS ACID

Nitrogen gas (N_2) is evolved when an amino acid is treated with nitrous acid. Use 5 ml of glycine solution, solid sodium nitrite and a few drops of dilute hydrochloric acid. As a control, repeat the experiment using water instead of glycine solution.

3. NINHYDRIN REACTION

α-amino acids react with ninhydrin at pH >4 according to the scheme.

$$\text{R·CH(NH}_2\text{)COOH} + \text{ninhydrin·CH(OH)}_2 \longrightarrow \text{ninhydrin·CHOH} + \text{R·CHO} + NH_3 + CO_2$$

α-amino acid   ninhydrin   coloured complex

The reduction product of ninhydrin then reacts with ammonia (NH$_3$) and excess ninhydrin to yield a blue coloured substance. The test is extremely sensitive, and care should be taken to avoid spilling ninhydrin solution on the hands as the stains which result are difficult to remove.

To 1 ml of glycine solution add a few drops of 2% ninhydrin solution and boil over a flame for 2 minutes. Allow to cool and observe the blue colour.

Place 1 small drop of glycine solution on a filter paper and dry by warming gently. Place on the same spot 1 drop of ninhydrin solution and dry again. A blue or purple colour will appear after a few moments.

Specific reactions

Pure amino acids are expensive so only small amounts of solutions should be used in the following tests.

1. XANTHOPROTEIC TEST—for amino acids with an aromatic side chain

To 1 ml of solution add 1 ml of concentrated nitric acid. Heat over a flame for 2 minutes and observe that the colour of the precipitate changes to yellow. Cool the solution under the tap and cautiously run in sufficient 40% sodium hydroxide to make the solution strongly alkaline. Again observe the change in colour of the precipitate.

Test solutions of tryptophan, phenylalanine and glycine.

2. MILLON'S TEST

This reaction is only given by phenolic groups. Tyrosine is the only common phenolic amino acid. A yellow precipitate is **not** a positive reaction but usually indicates that the solution is too alkaline.

Add a few drops of Millon's reagent to 2 ml of neutral or acid solution and warm the tube in a boiling water bath. A brick red colouration is a positive reaction.

Test solutions of phenylalanine and tyrosine.

3. LEAD SULPHIDE TEST

This is a specific test for cystine. Some of the sulphur in the molecule is converted to sodium sulphide by boiling with 40% sodium hydroxide. The sodium sulphide can be detected by the precipitation of lead sulphide from an alkaline solution.

Add a few drops of lead acetate solution to 1 ml of amino acid solution. A precipitate results which is re-dissolved by adding a little 40% sodium hydroxide solution. On boiling, the solution turns dark brown.

Test solutions of phenylalanine and cystine.

4. HOPKINS-COLE TEST

This is a specific test for tryptophan. The indole group of tryptophan reacts with glyoxylic acid in the presence of concentrated sulphuric acid to give a purple colour. Glacial acetic acid which has been exposed to the light always contains glyoxylic acid as an impurity.

To a few millilitres of glacial acetic acid add 1-2 drops of tryptophan solution. Pour 1-2 ml of concentrated sulphuric acid down the side of the sloping test tube to form a layer underneath the acetic acid. The development of a purple colour at the junction of the two layers is a positive reaction. Repeat with a solution of glycine.

TESTS FOR PROTEINS

For the following tests dried albumin dissolved in water can serve as a convenient source of material. Protein may also be prepared from plant material by the following method. Grind up about 10 g of dried pea seeds add 100 ml of water and allow the mixture to stand for 1 hour. Filter, and use the filtrate in the tests which follow.

1. XANTHOPROTEIC TEST—as above.

2. NINHYDRIN TEST—page 68.

3. MILLON'S TEST—as above.

4. HOPKINS-COLE REACTION—as above.

5. LEAD SULPHIDE REACTION—as above.

6. BIURET TEST

To 2 ml of solution to be tested add an excess of sodium hydroxide solution and then a few drops of 1% copper sulphate solution. A violet colour results.
Test solutions of glycine and tryptophan.

For the following reactions, a protein free from other impurities is required, so that the solution of albumin should be used.

7. NEUTRAL SALTS

Ammonium sulphate is commonly used for the precipitation of different types of proteins. The concentration of the salt required to cause precipitation varies with the nature of the protein and the pH of the solution.

Add solid ammonium sulphate to about 5 ml of protein solution in a test tube. The salt should be added in quantities of 1 g at a time and the solution gently agitated after each addition to dissolve the ammonium sulphate.

8. PRECIPITATION BY HEAVY METALS

Measure out a few millilitres of protein solution into 3 test tubes and add a little copper sulphate solution to the first, lead acetate solution to the second, and mercuric chloride solution to the third. In each case the protein is precipitated.

9. PRECIPITATION BY ALCOHOL

To a few millilitres of protein solution, add an excess of absolute alcohol. The protein will be precipitated.

Experiment 44
To determine the isoelectric point of a protein

Note: The isoelectric point of a protein is that pH at which the net effective electric charge on the molecule is zero. At the isoelectric point, proteins exhibit minimum solubility, conductivity, osmotic pressure and viscosity.

APPARATUS REQUIRED
casein
50 ml volumetric flask
distilled water at 40°C
1 N sodium hydroxide
1 N acetic acid
18 test tubes
test tube rack
10 ml pipette
5 ml pipette

METHOD OF PROCEDURE

Place 0·25 g of pure casein in a 50 ml flask and add 25 ml of distilled water at 40°C. Then add 5 ml of 1 N NaOH and agitate until the casein dissolves. Avoid frothing as far as possible. Quickly add 5 ml of 1N acetic acid and make up the volume to 50 ml with distilled water. The resulting solution should be faintly opalescent.

Label a series of tubes 1 to 9. Into tube 1 measure 3·2 ml of 1 N acetic acid and 6·8 ml of distilled water. Mix the 2 solutions well. Into each of the other 8 tubes place 5 ml of distilled water. Using a pipette transfer 5 ml of the acid mixture from tube 1 to tube 2. Mix thoroughly and then transfer 5 ml from tube 2 to tube 3. Repeat this procedure throughout the series of tubes. To bring the contents of each tube to the same volume discard 5 ml of the mixture in tube 9. If the pipetting is done accurately, the pH of the solutions in tubes 1 to 9 will be:— 3·5, 3·8, 4·1, 4·4, 4·7, 5·0, 5·3, 5·6, 5·9 respectively. Arrange a second series of 9 tubes behind those already set up and into each tube place 1 ml of the casein solution. Pour the first series of tubes into the corresponding tubes containing casein and mix rapidly. Note the immediate effect and the effect after standing for 30 minutes.

Express the results obtained in a table form indicating the degree of opalescence by +, ++, +++ etc. or degree of precipitation by p, pp, ppp etc.

TESTS FOR LIPIDS

A. Lecithin

1. PHYSICAL PROPERTIES

Test the solubility of lecithin in water, warm alcohol and ether.

2. EMULSIFYING ACTION

Place 5 ml of water in each of 2 test tubes. To one add 2-3 drops of oil and to the other 2-3 drops of lecithin in oil solution. Shake each tube well and compare the stabilities of the emulsions. Can you explain this property of the lecithin molecule?

B. Cholesterol

1. PHYSICAL PROPERTIES

Test solubility in water, acetone and alcohol. Make a solution in warm alcohol, allow to cool slowly and examine the crystals under the microscope.

2. LIEBERMANN-BURCHARDT TEST

Dissolve a small amount of cholesterol in 2 ml of **dry** chloroform in a **dry** test tube. Add 10 drops of acetic anhydride and 2 drops of concentrated sulphuric acid. Shake the tube and allow to stand. Note the colour.

It is essential that the chloroform should have been **dried** over anhydrous calcium chloride, or the experiment will not work.

Extraction of Lipids from plant material

Weigh out 50 g of linseed or castor oil seeds and grind in a mortar. Transfer to a stoppered container and add sufficient ether to cover the ground up seed. Replace the stopper and allow the mixture to stand for 2 to 6 hours. Filter off the ether into a flask, fit a condenser and distil off the ether over an electric heater. If a heater is not available distil from a water bath of boiling water after the flame has been extinguished. Do not use ether near a naked flame. When the bulk of the ether has been removed, pour the residue into an evaporating dish on a water bath and drive off the rest of the ether. Transfer some of the residue to test tubes and make the following tests.

1. Test solubilities in water, alcohol and chloroform.
2. Add a little 1% osmic acid—a black colour is formed. This test is used to detect fat in sections.

ENZYMES

Through the process of photosynthesis, the green plant is able to use light energy to build carbohydrates from carbon dioxide and water. Some of the carbohydrate is then degraded by the process of respiration, and the energy which results from this degradation is used by the plant to convert the remainder of the carbohydrate, either directly, or indirectly into all the various organic substances known to be present in plant tissues. Both the synthetic and degradative processes involve chemical transformations, and while a great deal is known about many of these transformations, others have still not been fully elucidated. From a study of those transformations which are known, it has become apparent that all the chemical reactions which occur in the living cell are catalysed by a group of compounds called enzymes. It has further become clear that the function of the enzymes is to speed up the rate at which the chemical transformations take place, since the reactions have still been found to occur in their absence but only very slowly. Each cell of a plant is thought to contain about one thousand different enzymes and the efficiency of the cell depends on the efficiency with which these enzymes work together.

All the enzymes which have so far been isolated and purified have been found to be proteins, and it is generally believed that all enzymes are protein although the reverse is certainly not true.

The enzymes seem to have relatively large molecules since, although it is difficult to determine accurately the exact molecular weight of an isolated enzyme, estimations indicate that it is never less than 10,000 and may even exceed 250,000.

Because enzymes are proteins they are liable to injury from such agents as heat, strong acids, or alkalis, organic solvents, ultra-violet light, and heavy metals such as silver and lead. All these agents are said to "denature" proteins because they alter the protein in such a way as to prevent it functioning normally. In many cases the alteration is irreversible. An example of irreversible thermal denaturation is the change brought about in the white of an egg as a result of being boiled.

Although certain enzymes appear to be pure protein and are able to function efficiently on their own, many others are unable to act in the absence of additional co-factors, which are usually referred to as **prosthetic groups** or **co-enzymes.** In many cases the prosthetic group is a single metallic ion, of a metal such as copper, zinc, iron, manganese, or magnesium. In others it is a complex organic molecule such as FAD or one of the B group of vitamins.

Enzymes resemble other catalysts in that they are effective in minute amounts, and are not altered or destroyed during the reaction. They differ from other catalysts, however, in that their catalytic activity is confined to a single reaction or group of closely related reactions. In other words an enzyme will react with a specific compound (its substrate) or group of compounds, whereas an inorganic catalyst such as platinum is active in a wide variety of different reactions. The actual degree of specificity shown by enzymes varies. If an enzyme will react with only one compound, it is said to show **absolute substrate specificity.** An example of an enzyme of this type is glucokinase, which catalyses the transfer of a phosphate group from ATP to glucose. If another sugar, such as galactose, is substituted for the glucose, the enzyme will not function, despite the fact that galactose is very closely related to glucose.

Most enzymes however are said to show a **group specificity** since they react with groups of compounds having the same general structure. The enzyme phosphoprotein phosphotase, for example, will hydrolyse any phosphoprotein to its protein and phosphate components, irrespective of which protein is present.

A third type of specificity is shown by those enzymes which react only with the D-isomer or the L-isomer of a compound but not with both. Such enzymes are said to exhibit **optical specificity.**

An enzyme does not have its catalytic properties evenly distributed over its molecular surface but has them confined to a relatively small number of regions. For obvious reasons, these regions are known as the **active centres** of the enzyme, and some enzymes, of which trypsin, the digestive enzyme, is an example, have only one active centre, despite having quite large molecular weights. The number of active sites on an enzyme is believed rarely to exceed three or four.

FACTORS AFFECTING ENZYMIC REACTIONS

a) Enzyme concentration

In any enzyme catalysed reaction the rate at which the reaction proceeds is directly dependent on the amount of the enzyme present, provided that the substrate is present in amounts which ensure that the reaction is not held up by lack of material on which the enzyme can work. The more enzyme present, then the faster the reaction will proceed, since there will be more active centres available.

The relationship between the amount of enzyme present and the rate at which the reaction proceeds is shown diagrammatically below.

Graph showing the effect of enzyme concentration on the rate of reaction

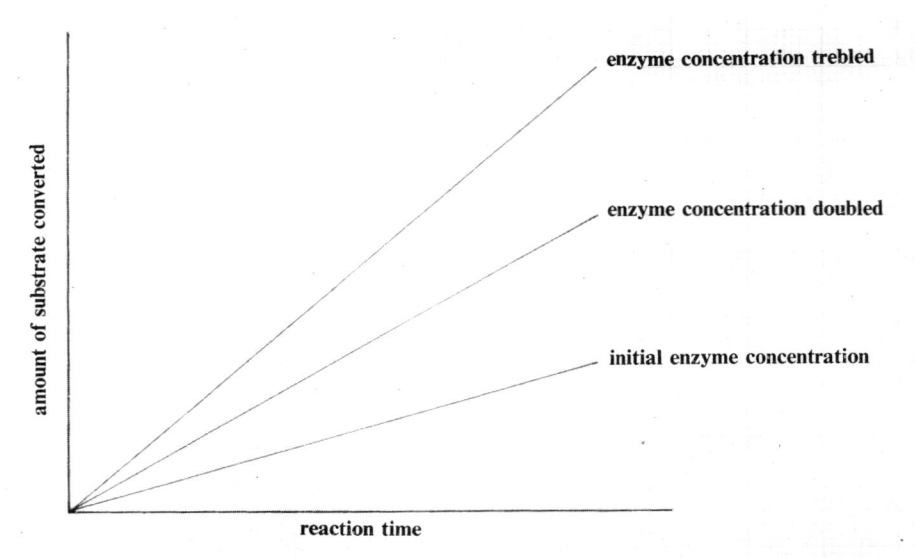

b) **Substrate concentration**

If the enzyme concentration is kept at a constant value and the substrate concentration increased from zero, then on plotting the rate of the reaction, against the concentration of the substrate, a characteristic curve is obtained. (see diagram below).

Graph showing the effect of substrate concentration on the rate of reaction

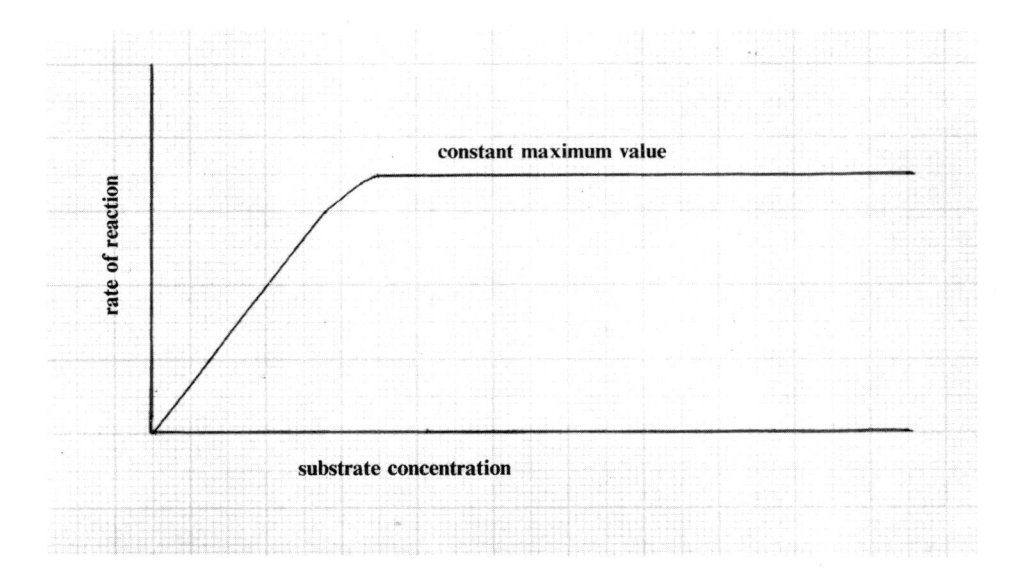

It can be seen that at first the curve rises linearly, then begins to flatten out and finally attains a constant maximum value. Over the linear part of the curve the enzyme is not working at its full capacity since there is not enough substrate to occupy all the active sites of the enzyme molecules. At these concentrations of substrate, doubling the amount of substrate present would double the rate at which the reaction occurs. Where the curve begins to flatten the enzyme is working at almost its maximum rate and most of the active centres are occupied. Doubling the substrate concentration here, would not result in the doubling of the reaction rate. Where the curve reaches its constant maximum value, the enzyme is working with maximum efficiency and all the active centres are in use. Increasing the amount of substrate present would have no effect on the rate of the reaction since there are no more active sites available for the extra substrate. The situation is analogous to that of a bricklayer, who can lay 500 bricks an hour. If he is supplied with only 100 bricks per hour then he can only build slowly, and giving him 200 bricks would allow him to double his building rate. Once the bricklayer is given 500 bricks an hour he is working as fast as he can and increasing the bricks to 600 per hour would only result in 100 bricks remaining unused each hour.

c) **Effect of temperature**

Like all chemical reactions, enzyme catalysed reactions are sensitive to changes in temperature and in general an increase in temperature results in an increase in the rate of the reaction. If however, temperatures above 45°C are used then there is an abrupt decrease in the activity of the enzyme because at these higher temperatures the enzyme becomes increasingly denaturised, which of course prevents it from functioning properly, if at all. Over the medium range of temperatures, enzyme catalysed reactions have Q_{10}'s approaching 2, (see experiment 52 on page 79).

d) **Effect of pH**

Enzymes in general are very sensitive to changes in pH and the result obtained on plotting the efficiency with which different enzymes work at various pH's is shown in the diagram below.

Graph showing the effect of pH on enzyme activity

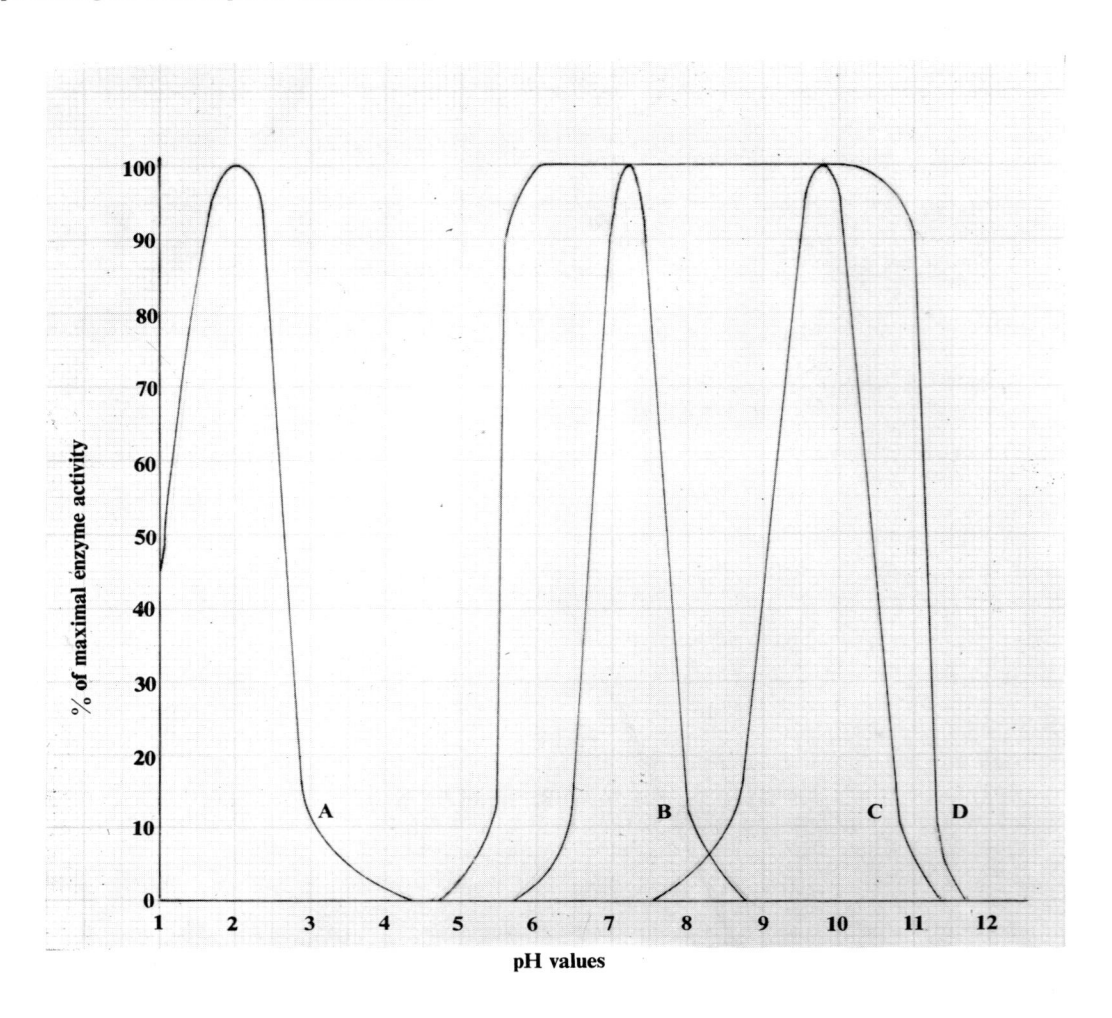

[Graph: % of maximal enzyme activity (y-axis, 0-100) vs pH values (x-axis, 1-12), showing four curves labelled A, B, C, D]

A. pepsin, optimum ph 2·0

B. diphosphoglyceromutase, optimum pH 7·2

C. glucose dehydrogenase, optimum pH 9·8

D. guanase, optimum pH 6 to 10

From the graph it can be seen that there is a pH range where the enzyme works with peak efficiency and that on each side of this range the efficiency of the enzyme decreases rapidly. The pH range over which the enzyme shows its greatest activity is referred to as its **pH optimum** and for most enzymes the pH optimum is narrow. Some enzymes, however, are able to work efficiently over a wide pH range. For example guanase, the enzyme which breaks down guanine, has an optimum pH range of from 6 to 10.

It can also be seen from the diagram above that different enzymes show their optimum activity at different pH values.

e) Inhibitors

Compounds which prevent enzymes from functioning properly are termed **inhibitors** and are generally considered to fall into two categories, namely **competitive** and **non-competitive inhibitors.**

COMPETITIVE INHIBITORS

These compounds usually have structures which are similar to that of the normal enzyme substrate. The enzyme is deceived into reacting with the inhibitor so that active centres which would normally be available to the correct substrate are occupied. The result is that the rate of attack on the proper substrate slows down. Competitive inhibition of an enzyme is easily overcome by increasing the amount of "correct" substrate available to the enzyme. The more substrate available, the more chance there is that an active site will combine with a molecule of the substrate and not with one of the inhibitor.

NON-COMPETITIVE INHIBITORS

Whereas the competitive inhibitor forms a loose attachment during its reaction with the enzyme, the non-competitive inhibitor forms a very strong attachment and is not displaced by increasing the substrate concentration. Many metals such as arsenic, mercury, and silver are non-competitive inhibitors of enzymes, which really explains why these metals are toxic to human beings and most animals. The degree to which a given enzyme is inhibited by these agents depends first of all on the ease with which the inhibitor combines with the enzyme, and secondly on the concentration of the inhibitor.

ENZYME EXPERIMENTS

Experiment 45

To demonstrate the action of the enzyme phosphorylase

Note: The enzyme phosphorylase catalyses the reaction:—

$$n \text{ glucose-1-phosphate} \rightleftharpoons \text{starch} + n \text{ (phosphate)}$$

with the equilibrium position depending on the reaction conditions. An active source of this enzyme is pea root tips.

APPARATUS REQUIRED
distilled water
citrate buffer pH 6·0
toluene
glucose-1-phosphate 0·4 M
4 test tubes in stand
3 × 1 ml pipettes
microscope slide
microscope
newly germinated pea seedlings

METHOD OF PROCEDURE
Set up 4 test tubes as shown in the table below.

Test tube	1	2	3	4
distilled water	1 ml	—	—	1 ml
citrate buffer pH 6·0	1 ml	1 ml	1 ml	1 ml
glucose-1-phosphate 0·4 M	—	1 ml	1 ml	—
toluene	—	—	2-3 drops	2-3 drops

Into each test tube, place a pea root tip. Squash another tip in iodine and examine for starch under the microscope. Leave the root tips in the 4 test tubes for as long as possible (at least 3 hours) and then test each for starch. Compare the relative amounts of starch present in tubes 2 and 3. Interpret your results.

Note: Toluene is a lipid solvent which destroys the structure of protoplasmic membranes and therefore allows substances to penetrate cells more easily.

Experiment 46
To demonstrate the action of the enzyme catalase

Note: This enzyme decomposes hydrogen peroxide:—

$$2 H_2O_2 \rightarrow 2H_2O + O_2$$

APPARATUS REQUIRED
potato
swede
dilute hydrogen peroxide
acid washed sand
pestle and mortar
distilled water
filter funnel
filter paper
4 test tubes in stand
2 × 1 ml graduated pipettes
2 × 5 ml graduated pipettes

METHOD OF PROCEDURE
Prepare extracts of a) potato and b) swede by grinding 10-15 grams of finely sliced tissue with sand in a pestle and mortar. Add 20 ml of distilled water, mix thoroughly and filter into an empty test tube. Test each extract for catalase activity by adding 0·1 ml of extract to 2 ml dilute hydrogen peroxide. The presence of the enzyme will be indicated by the production of bubbles of oxygen.

Compare the activity of the two extracts.

Experiment 47
To demonstrate the localisation of the dehydrogenase enzymes with tetrazolium staining

Note: Many dehydrogenases, for example malic dehydrogenase, and alcohol dehydrogenase, can utilise tetrazolium dyes (2.3.5 triphenyl tetrazolium chloride, T.T.C.) as electron acceptors, which then become reduced. Since the oxidised dye is colourless and the reduced dye is red, T.T.C. can be used to detect the presence of dehydrogenases.

APPARATUS REQUIRED
0·05% 2.3.5 triphenyl tetrazolium chloride (T.T.C.)
water
pea or bean seeds (part a)
pea seedlings   (part b)
knife or scalpel
2 petri dishes

a) **Localisation of dehydrogenase enzymes in seeds**

METHOD OF PROCEDURE
Soak a number of suitable seeds (pea or bean) overnight in water. Bisect these longitudinally and place them cut side down in a petri dish containing 10-20 ml of 0·05% T.T.C. After 2 hours examine and note the parts of the seeds which are stained.

b) **Localisation of dehydrogenase enzymes in intact plants**

METHOD OF PROCEDURE
Place an intact seedling with its roots in 0·05% T.T.C., also cut off the roots of another under water and put the cut ends in the solution. Examine after 2-3 hours and again the next day. Can you see any red stain in the seedlings? Try to locate the intracellular sites of the stain under the microscope.

Experiment 48

To demonstrate the inhibition of polyphenol oxidase

Note: This enzyme catalyses the oxidation of polyphenols to products which polymerise to form brown compounds. Most people will have at some time or other seen this enzyme in action since it is responsible for the discolouration of peeled apples and potatoes. The enzyme can be inhibited by sodium dithionite which is a strong reducing agent and prevents the oxidation process.

APPARATUS REQUIRED
potato
sand (acid washed)
mortar and pestle
filter paper
filter funnel
sodium dithionite
3 test tubes
test tube stand

METHOD OF PROCEDURE
Prepare an extract of potato in the way described in experiment 46 on page 76. Pour half of the extract into a test tube in which a few grains of sodium dithionite have been placed. Leave the tubes standing and compare the colours of the untreated and dithionite treated portions after 5, 30, and 60 minutes.

Experiment 49

To demonstrate the requirement of a prosthetic group by polyphenol oxidase

Note: This enzyme requires the presence of copper ions for activity. When copper is removed from the enzyme, for example by the chemical Dieca, the enzyme is inactivated. Catechol can be used as the enzyme substrate.

APPARATUS REQUIRED
10^{-3} M Dieca 40
distilled water
0·18M Catechol
potato
pestle and mortar
filter paper
filter funnel
acid washed sand
phosphate buffer pH 6·8
muslin
2 test tubes and stand
knife

METHOD OF PROCEDURE
Set up the test tubes with chemicals as shown in the table below.

test tube	1	2
distilled water	0·4 ml	0·5 ml
Dieca 40	0·1 ml	—
potato extract	3 ml	3 ml
Catechol	1 ml	1 ml

When these are ready, prepare the potato extract in the following way. Grind 15 grams of finely sliced potato tissue with sand and 15 ml buffer pH 6·8. Squeeze through muslin and **immediately** add to the tubes the required amount. Observe the colour at intervals.

Note: Do not get Catechol on your skin.

Experiment 50

To demonstrate the requirement of a prosthetic group by catalase

Note: This enzyme requires iron for its activity and may be inactivated by sodium azide which forms a complex with iron and renders it unavailable to the enzyme.

APPARATUS REQUIRED
potato
acid washed sand
mortar and pestle
filter funnel
filter paper
dilute hydrogen peroxide
0·5M sodium azide
4 test tubes in stand
knife

METHOD OF PROCEDURE
Prepare the enzyme extract as previously described in experiment 46 on page 76. Divide the extract into 2 portions. Add one portion to a test tube containing 5 ml of dilute peroxide solution, and the other portion to a test tube containing 4 ml of peroxide and 1 ml of 0·5M sodium azide. Compare the rate of oxygen evolution in the two tubes.

Experiment 51

To determine the pH optima of an enzyme using amylase

Note: The enzyme used is amylase which catalyses the hydrolysis of starch to glucose.

$$(C_6 H_{10} O_5)n + n H_2O \rightarrow n C_6 H_{12} O_6$$

The enzyme can be obtained commercially, in which case a 0·2% solution must be prepared from the concentrate, or ordinary saliva may be used as a source of the enzyme in which case a 100 times dilution is necessary.

APPARATUS REQUIRED
amylase enzyme—as above
phosphate buffers ranging from pH 6·0 to pH 8·0
distilled water
potassium dihydrogen phosphate (0·5M)
disodium hydrogen phosphate (0·5M)
9 test tubes
0·1% solution of starch
iodine solution
7 glass rods
white tile
graph paper
test tube stand

METHOD OF PROCEDURE
a) Preparation of the buffers
The buffers are prepared by mixing together various amounts of 0·5M potassium dihydrogen phosphate, 0·5M disodium hydrogen phosphate and distilled water. The actual amounts of each are shown in the table below where A = potassium dihydrogen phosphate and B = disodium hydrogen phosphate.

pH	A	B	Distilled water
6·0	74·2 ml	8·5 ml	17·3 ml
6·4	53·4 ml	15·5 ml	31·1 ml
6·8	31·4 ml	22·8 ml	45·8 ml
7·0	22·4 ml	25·8 ml	51·8 ml
7·2	15·4 ml	28·2 ml	56·4 ml
7·6	6·7 ml	31·0 ml	62·3 ml
8·0	2·8 ml	32·4 ml	64·8 ml

Place 5 ml portions of the buffers, one in each test tube, and then add to each tube 5 ml of a 0·1% solution of starch and 5 ml of the amylase solution. Mix thoroughly. Place drops of iodine on a white tile and, using a different glass rod for each tube, transfer a drop of each solution to a drop of iodine. Continue sampling from the reaction mixtures at 2 minute intervals until the starch has been digested. Take care to clean the rods after each sample.

Note the time taken in each case for the digestion of the starch. Draw a graph in which the efficiency of the enzyme, 1/time in minutes, is plotted against pH. From the graph determine the optima pH of the enzyme.

Experiment 52
To determine the effect of temperature on an enzyme, using amylase

APPARATUS REQUIRED
6 test tubes in stand
0·1% starch solution
phosphate buffer
amylase solution
iodine
white tile
6 glass rods
graph paper
6 water baths, 4 of which should be heated in some way; all 6 should have thermometers resting in them to ensure correct regulation of the temperature

METHOD OF PROCEDURE
6 test tubes are set up, each containing 5 ml of 0·1% starch solution and 5 ml of the buffer of the pH at which the amylase showed its optimum activity (see experiment 51). These are placed one in each of the water baths which are at approximately 10°C (running tap water), approximately 20-22°C (room temperature) 30°C, 40°C, 50°C and 60°C or over. To each tube add 5 ml of the 0·2% amylase (or 5 ml of saliva water) and mix. At 2 minute intervals test for the presence of starch in the manner described in experiment 51. At each temperature note the time taken for the complete disappearance of the starch. Plot your results as a graph showing activity (1/time in minutes) against temperature. At what temperature does the enzyme show its optimum activity?

To determine Q_{10} for an enzyme reaction divide the rate at $(t + 10)°C$ by the rate at $t°C$:—

$$Q_{10} = \frac{\text{rate at } (t + 10)°C}{\text{rate at } t°C}$$

since the rate is the inverse of the time taken then,

$$Q_{10} = \frac{\text{time taken at } t°C}{\text{time taken at } (t + 10)°C}$$

Experiment 53
To demonstrate the action of the enzyme urease

Note: The enzyme urease catalyses the hydrolysis of urea:—

$$NH_2 CONH_2 + H_2O \longrightarrow 2NH_3 + CO_2$$

APPARATUS REQUIRED
urease solution (commercially available)
1% urea solution
phenol red solution
dilute acetic acid
dilute sodium hydroxide
water bath at 37°C
5 ml pipette
2 pasteur pipettes
2 test tubes in stand

METHOD OF PROCEDURE

Before starting the experiment, familiarise yourself with the colour of the indicator phenol red in acid and basic solutions. This can be done by adding a few drops to dilute acetic acid and also to dilute sodium hydroxide.

Into each test tube pipette 5 ml of 1% urea solution and 5 drops of phenol red. If necessary, adjust each mixture to about pH7 (orange-yellow) with a drop of dilute acetic acid. To one tube add 5 drops of urease solution and to the second tube 5 drops of boiled urease solution. Place both tubes in a water bath at 37°C. Note the colour change of the indicator as the enzyme hydrolyses the urea. Explain why the action of the enzyme causes this change.

Experiment 54
To demonstrate, using the Thunberg technique, the presence of a dehydrogenase enzyme in milk

Note: Dehydrogenase enzymes oxidise their substrate by removing hydrogen from it. The hydrogen is passed on to a suitable acceptor, such as NAD or NADP, and ultimately via the electron transport chain to atmospheric oxygen.

In the Thunberg technique the process is carried out in the absence of oxygen and the hydrogen is transferred to an artificial acceptor such as methylene blue. The latter is reduced to a colourless leuco form as a result of accepting the hydrogen.

APPARATUS REQUIRED
2 Thunberg tubes
water pumps
0·01% methylene blue
distilled water
10 ml fresh, non-pasteurised milk
0·2M phosphate buffer pH 6·8
saturated solution of benzaldehyde
short length of rubber tubing
white petroleum jelly
5 ml pipette
3 × 1 ml pipette

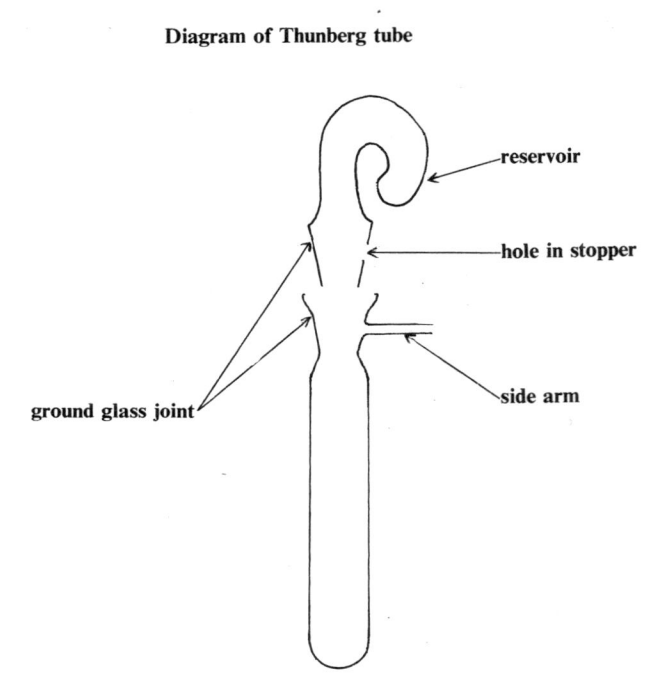

Diagram of Thunberg tube

(labels: reservoir, hole in stopper, side arm, ground glass joint)

METHOD OF PROCEDURE

Grease the stoppers of both of the Thunberg tubes and in the body of the first, place 1 ml of phosphate buffer, 1 ml of 0·01% methylene blue, 1 ml of water and 2 ml of fresh milk. As a control set up the second tube in the same way but use boiled milk instead of fresh milk. Into the hollow stopper of each tube, pipette 1 ml of the saturated benzaldehyde solution. This is the substrate for the enzyme.

Carefully replace the stoppers in the tubes and rotate them to give a good airtight seal. Make sure the stoppers are left in the open position with the interior of the tube connecting with the atmosphere through the side arm. Evacuate each tube in turn by connecting the side-arm through a short length of rubber tubing to a water pump. The evacuation is effected most efficiently by holding the tube at an angle of about 30-35°, thus exposing as large a surface area of the solutions as possible without allowing them to mix. During the evacuation tap the tube **gently** on the bench at intervals to facilitate the removal of dissolved air. Evacuate each tube for at least 5 minutes, and then rotate the stopper so that the tube is sealed. Once the tube is sealed, disconnect it from the water pump.

When both tubes are ready, tip the contents of the stoppers into the bodies. The action of the enzyme results in the dye becoming decolourised as it becomes reduced as a result of accepting hydrogen molecules.

At the end of the experiment rotate the stoppers back to the open position and note what happens when air is re-admitted to the tubes. Why does this happen?

CHROMATOGRAPHY

The naturally occurring substances found in biological systems are usually present as complex mixtures. The separation of these compounds, one from another, in order to study their activities and relative importances, proved to be very difficult using only the classical methods of the chemist, and it was not until the development of chromatographic techniques that major advances in the study of plant and animal metabolism became possible.

Chromatography is a method of separation, and the advantages of using chromatographic separation techniques for analytical purposes lies firstly, in their ability to separate and isolate the constituents of complicated mixtures of chemically related substances, and secondly in their great sensitivity which means that they can be used to detect very small quantities of the various compounds. These two properties make them ideally suited for use with plant and animal extracts in which the most active compounds are very often present in minute amounts.

Chromatography was first discovered by Twsett as long ago as 1903, when he found that he could separate the various plant pigments from each other by passing a solution containing them, down a vertical column of calcium carbonate. By this means he showed that there were two types of chlorophyll (chlorophyll a and chlorophyll b) present in higher plants, a fact which had long been suspected but which until then had remained unproven, due to the inability of workers to separate them from each other. Twsett also demonstrated that, in addition to the chlorophylls, there are two other groups of pigments present, namely the carotenes and the xanthophylls. His discovery that mixtures containing several compounds could be separated into their constituents in this way remained un-noticed until the 1930's when research workers at last realised its potential, and began to develop more refined techniques based on his original findings.

At first the techniques were rather clumsy and could only be used to separate mixtures containing a few types of compounds, but, as the techniques improved, so the range of compounds which could be separated by chromatography was extended, until today it includes most types of organic compounds, for example: amino acids, nucleic acids, carbohydrates, pigments, lipids, sugar acids, proteins etc.

Similarly, the techniques at first served only as a means of determining whether or not a substance was present in a mixture, but gave little information as to the actual amounts of that substance present. Modern chromatographic techniques, however, allow the quantitative determination of amounts as low as 1-3 μ grams.

Although originally designed for investigation into the biochemical nature of plant and animal life, chromatography has made valuable contributions to other fields, particularly those of pharmaceutics and medicine where it is used extensively for the isolation and purification of drugs and antibiotics, and also in many cases as a means of diagnosing or confirming a particular illness or condition. For example, abnormal carbohydrate metabolism, which leads to the appearance in the urine of unusually large amounts of sugar, can be detected easily and quickly by a chromatographic analysis of urine samples. Forensic science has also benefited greatly from the use of chromatographic techniques since they have made it easier to detect dopes and poisons used on animals or human beings.

It can safely be said that virtually no biological or chemical field has failed to benefit from the development of these techniques.

The chromatographic techniques used today can be divided into three general types depending on whether they use the properties of adsorption, partition or ion-exchange to bring about the separation of the constituents of complex mixtures. Although the method by which the separation is achieved may differ, all three types have certain common features when compared with the old techniques of precipitation, sedimentation, crystallisation etc. They are all simpler and quicker to use, have a higher efficiency of separation and can be successfully used with very much smaller amounts of material.

The method of procedure is basically similar for all three types of chromatography and may be conveniently divided into six stages.

1. The preparation of the sample for analysis.
2. The preparation of the chromatogram.
3. The application of the sample to the chromatogram.
4. The choice of the solvent which brings about the separation.
5. The actual separation of the sample into its constituents (often referred to as the "development" or "running" of the chromatogram).
6. The identification of the separated compounds.

Since the preparation of the sample is carried out in the same way, irrespective of the method used for its subsequent analysis, the procedure may be discussed at this point. The remaining five stages, however, vary from method to method and are best described in the appropriate sections.

Preparation of the sample for analysis
The material for analysis must be dissolved in a suitable solvent, the nature of which depends on the type of compounds being investigated. Solvents commonly used include hot or cold aqueous alcohol, ethyl acetate, trichloroacetic acid, chloroform, ether, benzine, acetone and water. Where the investigation is one into the nature of compounds present in an organ such as a leaf, or root, the organ is ground or macerated in a few millilitres of the solvent and left to stand for a short period. The mixture is filtered and the filtrate, which contains a variety of dissolved compounds, is then either applied directly to the chromatogram or is first concentrated before being applied. The use of organic solvents facilitates the concentration of the solvent.

ADSORPTION CHROMATOGRAPHY

When a mixture of substances dissolved in a liquid (phase 1), is brought into contact with a solid (phase 2), then the substances tend to accumulate at the interface between the two phases. This process is called **adsorption** and the fact that not all substances are adsorbed to the same degree has been utilised in the development of this particular type of chromatography.

PREPARATION OF THE CHROMATOGRAM
Adsorption chromatography is usually carried out on a vertical column of an adsorbent material. The most commonly used adsorbents are dehydrated aluminium oxide, activated charcoal, calcium carbonate, silica gel, zinc carbonate, calcium phosphate and sugar. All these compounds are commercially available in forms which have been specially prepared for chromatography to ensure uniform grain size, a property which is essential for the preparation of a good column free from large air spaces.

The column itself is prepared by introducing a suitable amount of the adsorbent into a glass tube, the diameter and length of which depends on the amount of material being analysed. The larger the amount of material being analysed, the larger and wider the tube needs to be. The lower end of the tube is fitted with a filter plate, or some glass wool, in order to retain the adsorbent, and a bung through which passes a short piece of fine bore glass tubing which allows the escape of the solvent after it has passed down the column. The glass tubing is itself connected to a piece of rubber tubing which has a clip attached to it, thus allowing the regulation of the rate at which the solvent escapes.

The adsorbent material may either be introduced into the glass tube as a dry powder and then the solvent, which is going to be used to effect separation added, or it may be washed in with the solvent. The latter method produces columns which are looser but more regularly packed. In the preparation of the chromatogram it is essential to avoid packing the column too tightly, as this leads to blockages during the development. The top of the column should always be covered by the solvent.

APPLICATION OF THE SAMPLE TO THE CHROMATOGRAM
In adsorption chromatography, the application of the sample to the column and the choice of the solvent which brings about the separation are inter-related, since the sample must be introduced onto the column dissolved in a small amount of this solvent.

Immediately before the application of the sample to the chromatogram, most of the excess solvent is removed from above the column to prevent undue dilution of the sample itself. The sample is then introduced, with a pipette, to the top of the column, care being taken to avoid disturbing the surface of the column unduly. The surface can be protected by a piece of filter paper, cut to the appropriate size.

DEVELOPMENT OF THE CHROMATOGRAM
The clip at the bottom of the column is opened slightly, allowing the escape of some of the solvent, and causing the sample to pass into the upper end of the column. When almost all the sample has entered, sufficient solvent is added to the top of the column to ensure a continuous flow through the column during the period of development. The actual rate of flow of the solvent can be regulated by adjustment of the clip at the lower end of the column. In general, the lower the rate of flow, the sharper the separation of the constituents of the sample being analysed. However, a compromise is often necessary owing to the time available for the analysis.

As the sample is carried through the column by the flow of the solvent its constituents become separated from one another as a result of their being adsorbed differentially onto the column. The strength with which a compound is adsorbed may be regarded as its resistance to being washed through the column. The most weakly adsorbed compound will have little resistance and will pass through the column most quickly. The most strongly adsorbed compound, however, offers the greatest resistance and will therefore pass down the column of adsorbent most slowly. Between those two extremes there will be a number of compounds showing various strengths of adsorption, that is, various degrees of resistance and therefore travelling at various speeds.

In those instances where the substances being separated are coloured it is a simple task to decide when they have been separated enough to allow identification. When the substances present in the sample are colourless then it is more difficult, and generally the decision as to when separation is complete is based on previous experience since it has been found that under standard conditions two or more substances will always take the same amount of time to separate.

The separated compounds may be collected in separate test tubes as they are washed out of the bottom of the column, or the development may be stopped before the fastest moving compound leaves the column, simply by closing the clip at the lower end. Generally the latter procedure is more usual where the compounds being separated are coloured and the former when they are colourless.

IDENTIFICATION OF THE SEPARATED COMPOUNDS

When the substances being separated are coloured and development is stopped before they leave the column, then identification of each compound is based on its colours and also its position in the column. When the substances are colourless, and are washed from the column into separate tubes then each tube must be tested individually with a reagent. The collection of the eluate into a series of tubes, as it leaves the column, requires an automatic collecting apparatus which ensures that each tube collects the same amount. The cost of this apparatus may be prohibitive to many schools and should it not be available then development of the chromatogram may be stopped before the compounds begin to leave the column. The identification of the separated, colourless compounds can then be achieved by carefully pushing the moist column out of the tube with a suitable implement and then applying, with a fine brush, a reagent which reacts with the various substances to produce characteristic colours.

PARTITION CHROMATOGRAPHY

When a substance is dissolved in a solvent such as water and the water added to an organic solvent with which it is immiscible, then some of the substance will pass from the water into the organic solvent, that is the substance becomes distributed between the water and the organic solvent. If, for instance, carbon tetrachloride is added to an aqueous solution of iodine, then some of the iodine passes into the tetrachloride. This can be easily proved by testing a sample of the carbon tetrachloride with a little starch solution.

The ratio, of the concentration of the substance in the water, to the concentration of the substance in the organic solvent, is a constant value (K_c) which is known as the **distribution coefficient,** or **coefficient of partition.**

$$\frac{\text{concentration in solvent 1}}{\text{concentration in solvent 2}} = K_c.$$

Partition chromatography utilises the fact that different substances have different coefficients of partition, to bring about the separation of these substances from one another.

The chromatography is always carried out on filter paper (paper chromatography) or on thin layers of specially prepared materials spread on glass plates, and always involves the use of a separation solvent which is a mixture of water and one or more organic solutions. The paper has a very high capacity for holding water, so that the aqueous part of the solvent mixture remains more or less stationary in the paper, while the organic part of the solvent mixture moves through the paper, either by capillarity or in response to gravity, depending on which way the chromatogram is developed. The organic part of the solvent may be regarded as the force which pushes the compounds along the paper or the thin layer, while the partition of the substances between the mobile organic phase and the stationary water phase represents the brake which slows down their progress. If there were no partition then the sample would move along unseparated with the organic solvent. The speed with which a compound moves depends on its coefficient of partition and because each compound has its own characteristic partition coefficient, it is possible to separate them in this way.

Paper Chromatography

The separation of compounds by paper chromatography can be achieved either by allowing the solvent to pass, under the influence of gravity, downwards through the paper, a technique described for obvious reasons as **descending paper chromatography,** or by **ascending chromatography,** in which case the solvent rises through the paper by capillarity. The latter technique has the advantages that it needs only simple inexpensive equipment and is often quicker, but the disadvantage that it does not generally produce as effective a separation as can be achieved by the descending technique. The basic equipment required for both types of chromatography is the same, namely, pure filter papers, solvents, a container or tank, capillary pipettes with which to apply the samples to the chromatogram, and apparatus for applying a detection reagent. This may take the form of a spray applicator or a container into which the chromatogram may be dipped.

Descending Paper Chromatography

Because the chromatograms are developed in a tank or container which must be prepared to receive the chromatogram, the method of procedure differs slightly from the general procedure described on page 81 and is as follows:—

1. Preparation of the sample for analysis—as previously described on page 82.
2. Preparation of the papers.
3. Preparation of the tank.
4. Application of the samples to the paper.
5. Development of the chromatogram.
6. Detection and identification of the separated compounds.

PREPARATION OF THE PAPERS

A wide range of filter papers, specially prepared for chromatography, can be obtained from various commercial sources, each paper having its own characteristic flow rate, strength, and capacity of loading. Whatman No. 1 grade paper is particularly useful since it is easy to work with, and can be used for the separation of a wide variety of compounds. The paper is available in widths ranging from 5 cm to 58 cm. It may be purchased already cut to a definite length (57 cm or 68 cm) or may be obtained as a roll from which lengths may be cut. The paper is generally supplied in boxes carrying arrows indicating the direction in which the solvent should be allowed to flow along the paper. Should the solvent be "run" in the opposite direction then separations are not as clear because compounds will tend to streak instead of running as discreet zones.

A pencil line should be drawn across the paper 10 cm from the top as indicated in the diagram below. Along this line, the starting line, a series of pencil marks should be made indicating where the samples for analysis are to be placed. These marks should be a minimum of 3 cm apart and should begin not less than 4 cm from the edge of the paper. The region of the chromatogram behind the starting line will be used to fix the paper in the chromatography tank.

Diagram of chromatogram

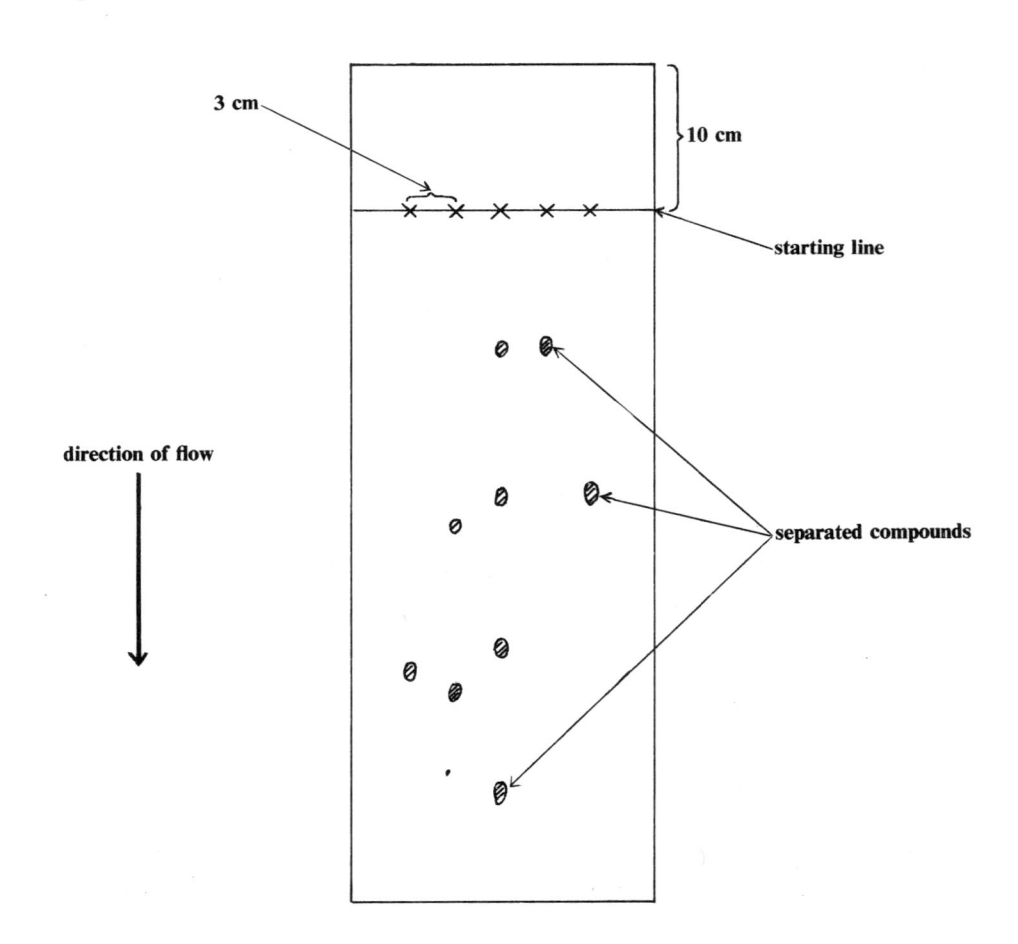

PREPARATION OF THE TANK

The container in which the chromatogram is developed is known as the chromatography tank and for descending chromatography the upper part of the tank must carry a glass trough in which solvent can be placed.

Suitable chromatography tanks are available commercially in a wide range of sizes. A useful sized tank, which has the added attraction of being fairly inexpensive, is the Shandon Unikit No. 1 Tank which accepts papers of widths up to 130 cm. This tank can also be used for ascending chromatography.

In order to cut down evaporation of the solvent from the surface of the paper during running, a phenomenon which can produce odd results, it is best to saturate the atmosphere of the tank with solvent vapour. This is done by placing a beaker containing some of the solvent in the bottom of the tank, some time before the chromatogram is introduced.

APPLICATION OF THE SAMPLES TO THE CHROMATOGRAPHY PAPER

The samples are applied to the paper using capillary pipettes. These are easily prepared by drawing out pieces of glass tubing in a bunsen flame, but, before use, the ends of the pipettes should be flamed briefly to ensure that they are not jagged or sharp enough to damage the paper. The end of a pipette is dipped into the required sample and the sample allowed to enter by capillarity. The pipette is then removed, held at an angle from the vertical and touched on the chromatogram at the appropriate mark on the starting line. The solution should then flow onto the paper. Care should be taken to ensure that the diameter of the resultant spot does not exceed 7-8 mm. The spot should be allowed to dry before further amounts of the sample are applied. Drying may be aided by a stream of cold air from a hair drier. The actual amounts of sample applied will vary according to the source and nature of the sample, but care must be taken to avoid placing too much on the chromatogram (overloading) as this results in poor separations. Where any doubt exists it is better to place a series of different amounts on different marks, since one or more of these should yield a good separation.

DEVELOPMENT OF THE CHROMATOGRAM

When the chromatogram is ready, it is placed in the tank suspended over one glass rod and held, at the top end in the trough, by another glass rod; see diagram below.

Diagrammatic representation of a descending chromatography tank with chromatogram

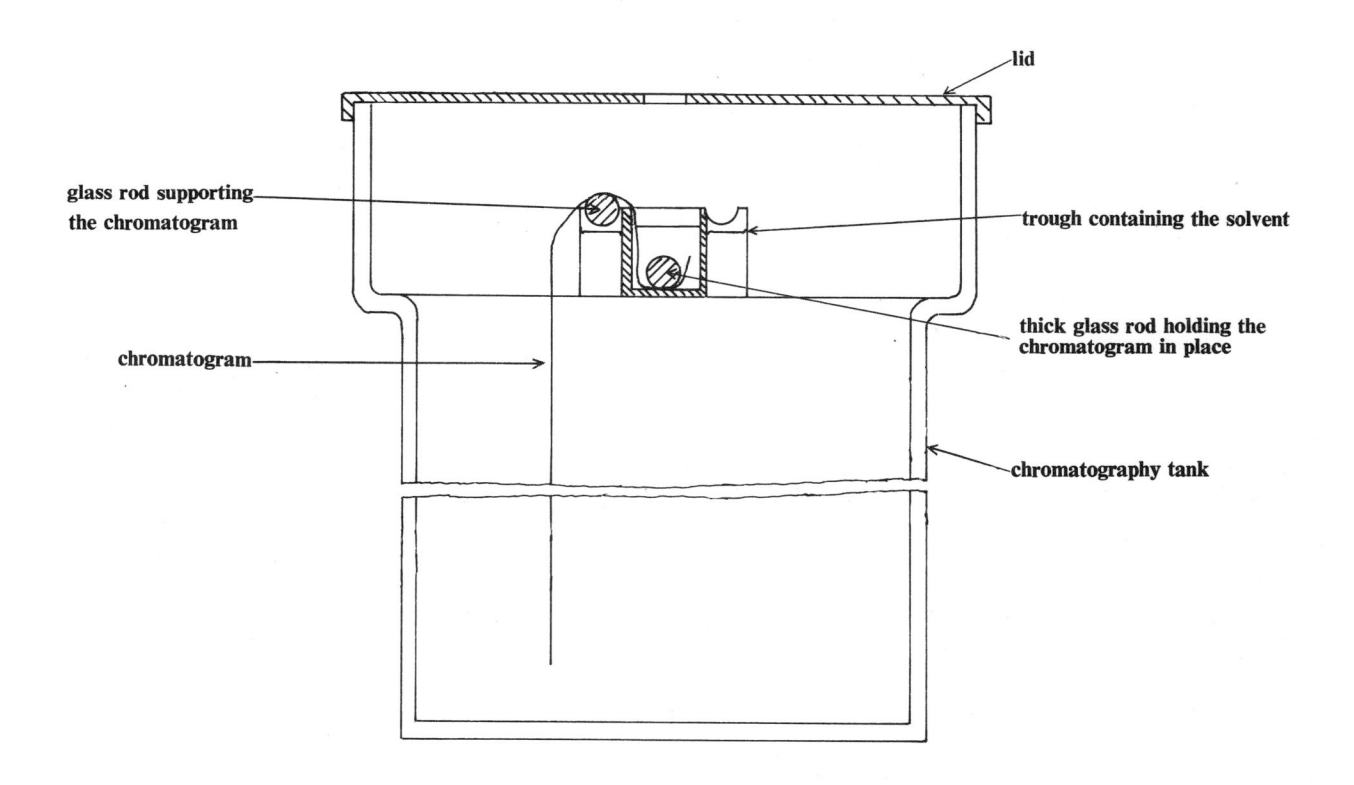

Sufficient solvent is then added to the trough to ensure that the chromatogram does not run dry during the development.

The type of solvent used depends on the type of compounds being separated. Relatively fast moving compounds such as amino-acids, can be separated by a slow running solvent, such as a mixture of butanol, acetic acid and water. For the separation of slow moving compounds such as sugars, a much faster running solvent such as a mixture of ethyl acetate, pyridine and water, must be used.

In order to prevent the escape of solvent vapour during the development of the chromatogram, the lid of the tank is sealed with white petroleum jelly, or high melting point paraffin wax. The chromatogram is then left to develop for a certain period. In the case of amino acids the development is stopped when the solvent front has run to within a few inches of the end of the paper. For sugars, the development time is much longer and is at least 18-24 hours. When development is completed, the chromatogram is removed from the tank, the position of the solvent front marked with a pencil, and the paper hung to dry at room temperature. Since many of the solvents used are organic and have obnoxious or irritating vapours, the drying is best done in a fume cupboard. When the chromatogram is dry to the touch, the separated compounds may be detected and identified.

DETECTION OF THE SEPARATED COMPOUNDS

As with adsorption chromatography, the results of the separation of the samples will only be visible if the substances being separated are coloured. Where this is not the case, a detection reagent, which reacts with the separated compounds to produce coloured derivatives, has to be applied to the chromatogram. In some cases, colour development will occur at room temperature, in others heat is required.

If an applicator is available then the reagent may be applied as a fine spray over the dried chromatogram. Spraying should be carried out in a fume cupboard and care should be taken to apply the spray evenly. BDH and Shandon now supply certain detection reagents in aerosol dispensers. In the event of such an applicator not being available then the chromatogram must be "dipped" into the reagent. In this case the detection reagent is placed in a shallow container and the chromatogram drawn through it in one continuous movement, see diagram below.

Diagram showing method of passing chromatogram through vessel containing detection reagent

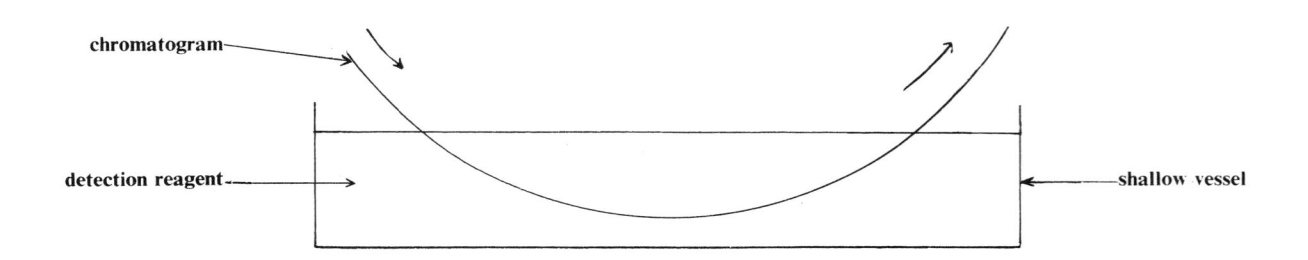

If the reagent is one which requires heat then after application of the reagent the chromatogram is placed in an oven at the desired temperature for a suitable period which varies with the reagent used. If no oven is available then a hotplate may be used instead. The separated compounds appear as coloured spots, the strength or intensity of which varies according to the amount of the compound which is present. The intensity of the colour of the spot can be used to determine the amount of the substance present, since over a certain range of concentrations, the intensity of the colour produced is proportional to the amount of substance reacting with the detection reagent. The intensity of the colour is first determined for a range of known concentrations and a graph prepared in which intensity is plotted against concentration. The intensity of the coloured spot originating from the sample is then determined and the concentration of the substance it represents, calculated from the graph. This method of determination is both sensitive and accurate and can be used to estimate very small amounts of material.

IDENTIFICATION OF THE SEPARATED COMPOUNDS

Many of the detection reagents are specific for certain types of compound and some produce characteristic colours with certain individual compounds. The ninhydrin reagent, for example, is specific for amino acids, yielding blue or purple colours, and the p-anisidine phosphate reagent is specific for sugars. This latter reagent is particularly useful since it yields characteristic colours with different types of sugars, hexose sugars appearing green-brown, pentoses appearing red and sugar acids appearing dark brown.

The position to which a compound moves during chromatography also gives a clue to its identity. The rate at which a compound moves on a chromatogram is known as its Rf value and is defined as follows:—

$$\text{Rate of flow (Rf)} = \frac{\text{the distance which the compound moves}}{\text{the distance which the solvent front moves}}$$

Each compound has a characteristic Rf value for each solvent. For example the amino acid glycine has an Rf value of 0·23 when run in the butanol: acetic acid: water solvent, and an Rf of 0·42 if the chromatogram is developed with an aqueous solution of phenol. If when analysing a sample of plant juice it was found that there was present a compound which had these Rf values in the two solvents, then this would be an indication that the compound could be glycine. The Rf value cannot be taken as the sole criterion for identity but is a very useful guide to the identity of a compound, particularly if used in conjunction with specific colour tests.

Ascending paper chromatography
One advantage of this type of paper chromatography is that it can be successfully carried out using very simple equipment, indeed any large glass container which can be made airtight can be used as the development tank. Large sweet jars are particularly useful in this respect.

The paper is prepared as previously described on page 84, but this time the starting line is drawn only 2 to 3 cm from the edge of the paper. This ensures that the sample spots are not immersed in the solvent and at the same time allows maximum development.

After the sample spots are dry, the paper is formed into a cylinder using linked paper clips, which prevents the edges from touching, see diagram below.

Diagram showing paper, with sample spots, folded ready for development

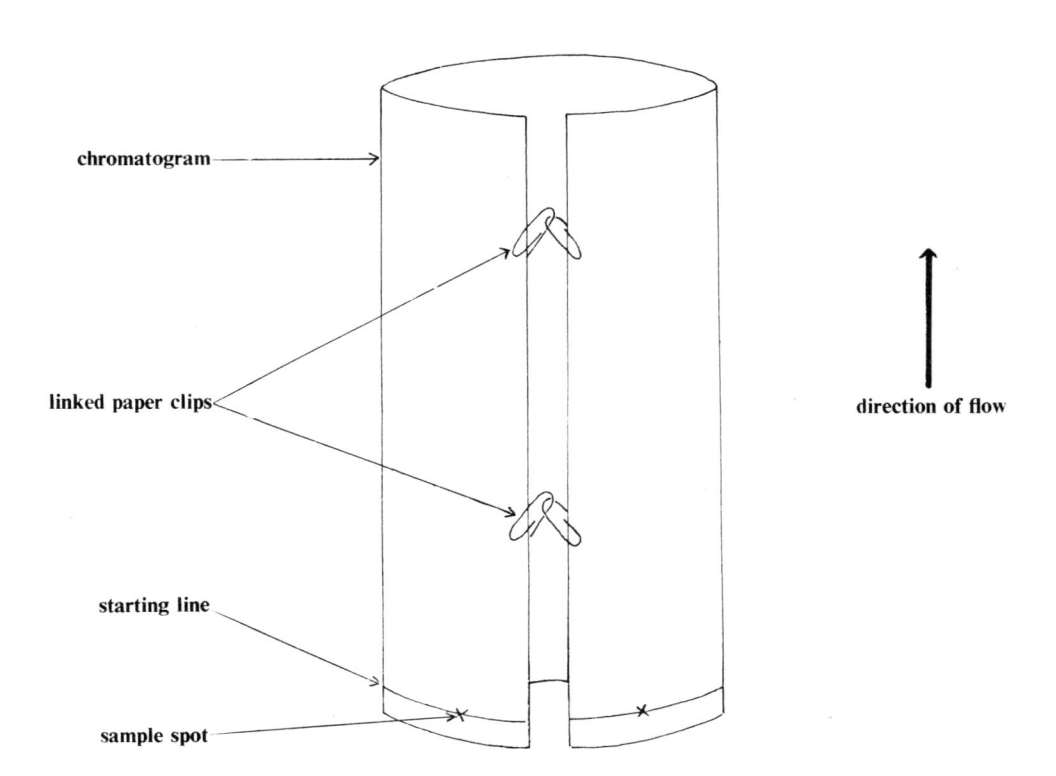

A suitable amount of solvent is added to the bottom of the tank or jar and the chromatogram stood upright in it with the sample spots lowermost. The solvent moves upwards by capillarity and development is complete when the solvent front reaches to within a few centimetres of the upper edge of the paper. The chromatogram is then removed, dried and if necessary the detection reagents applied (see page 86).

Ascending chromatography is unsuitable for the separation of sugars since it does not allow a sufficiently long development period.

THIN LAYER CHROMATOGRAPHY

Thin layer chromatography, a relatively recent development in the field of chromatography, provides a more effective method than paper chromatography for the separation of many types of compounds, since it is more sensitive and much quicker than the conventional methods of chromatography.

In the technique of thin layer chromatography a thin layer of a suitable adsorbent material, in the form of a paste, is spread over pieces of plate glass of appropriate size. Several types of material are now commercially available (Camlab Glass Ltd., Cambridge), but the most commonly used are silica gel and cellulose powders. The powders are specially prepared for thin layer work and have included with them a small percentage of "binding" material to assist in the formation of firm layers. The size of glass plate used varies from pieces the size of microscope slides to 20 cm squares.

After the layers have been spread and dried, they are used in more or less the same way as the ascending paper, There are however one or two differences.

APPLICATION OF THE SAMPLES TO THE CHROMATOGRAM

The method used to introduce the material to be separated onto the thin layer plate is basically similar to that used in paper chromatography. The main points of difference in applying this technique to thin layer plates are:—

1. No starting line is drawn on the plate as this disrupts the layer and interferes with the running of the chromatogram. The small hole inevitably made by the capillary pipette whilst spotting may be ignored.
2. The sample spot should be kept as small as possible. This is best achieved by only momentarily touching the layer with the pipette containing the sample. To avoid too great an outflow from the pipette it should be held at an angle of 45°, or less, from the plate. It is advisable to practise the application technique, using a small piece of filter paper, if there is sufficient sample available.
3. Since this method is infinitely more sensitive, it is easier to overload the chromatogram. Only very small amounts of material need be applied.

PREPARATION OF THE TANK

The solvent is added to the tank a short time before the chromatogram is introduced, and the lid replaced to allow the atmosphere of the tank to become saturated with solvent vapour.

DEVELOPMENT OF THE CHROMATOGRAM

The thin layer plate is placed in the tank with the sample spots lowermost, and the lid replaced. Development is a very rapid process and is generally completed within 30 minutes. Because development is so rapid a line is drawn across the plate a few centimetres from the upper end, before the chromatogram is placed in the tank. This disrupts the layer and prevents the solvent from passing any further, and therefore prevents the chromatogram from over-running. When development is complete, the plate is removed and dried in a stream of cold air.

DETECTION AND IDENTIFICATION OF THE SEPARATED COMPOUNDS

Thin layer plates cannot be dipped into the detection reagents because this washes off the layers. The reagents can only be applied in the form of a gentle spray.

Identification of the detected compounds is by their colour reactions and Rf values (see pages 86 and 87).

The apparatus needed for thin layer chromatography is both complicated and costly. However, it is possible to demonstrate thin layer chromatography in a simple form using a suitable layer, spread over a microscope slide, (see experiment 59 on page 93).

ION-EXCHANGE CHROMATOGRAPHY

This technique is outside the scope of this book, but basically it depends on the fact that substances forming ions in aqueous solution carry a certain electric charge. When a solution containing such substances is passed over an ion-exchange resin (material specially prepared for this type of chromatography), the resin binds ions of the opposite charge to the functional groups of the resin, and rejects those carrying the same charge. The strength of the binding depends, of course, on the magnitude of the charge and the size of the ion.

CHROMATOGRAPHIC EXPERIMENTS

The first two experiments are experiments in which chromatographic techniques are used to separate the constituents of a mixture of several purified compounds. The remaining experiments involve the separation of compounds extracted by various means from plant tissue.

Experiment 55

Experiment to demonstrate the separation of amino acids by ascending paper chromatography

APPARATUS REQUIRED

1 large sweet jar with a screw top (or other suitable container)
solutions of a. glycine
            b. alanine
            c. tryptophan   prepared by dissolving 10·20 mg of amino acid in 10 ml of 10% isopropanol
            d. valine
            e. leucine

Whatman No. 1 grade chromatography paper
paper clips
developing solvent:— n-butanol: acetic acid: water (12 ml: 3 ml: 5 ml)
6 capillary pipettes
petri dish and paper clips
detection reagent—ninhydrin/pyridine

METHOD OF PROCEDURE

Cut a piece of Whatman paper $9\frac{3}{4}$ inches long and 9 inches wide. Draw the starting line 1 inch from the end of the paper and mark along this line a series of fine crosses at $1\frac{1}{2}$ inch intervals, commencing 1 inch from the edge of the paper. Label these lightly in pencil A, B, C, D, E and F respectively. Using a separate pippette for each sample, spot the chromatogram on the crosses as follows:—

    A. 1 drop of glycine
    B. 1 drop of alanine
    C. 1 drop of tryptophan
    D. amino acid mixture (1 drop of solutions A, B, C, E and F mixed together).
    E. 1 drop of leucine
    F. 1 drop of valine

After use the samples may be stored in a cool place for 4-6 weeks.

When the spots are dry, form the paper into a cylinder as previously described on page 87.

Prepare the jar for chromatography by placing in it a petri dish. Make sure this petri dish is level and then half fill the dish with the chromatography solvent. Replace the lid of the jar and then leave it for 15 minutes to allow the air in the jar to become saturated. Then remove the lid and stand the chromatogram in the petri dish, making sure that the spots are at the bottom end of the paper cylinder. If the jar is too small to take the petri dish, stand the chromatogram directly on the bottom of the jar. In this case however, it will be necessary to use twice the volume of solvent. Replace the screw top on the jar and allow the chromatogram to develop for about 4 hours. At the end of this time remove the chromatogram from the jar, mark the position of the solvent front and then dry the chromatogram at room temperature.

DETECTION OF THE AMINO ACIDS

The detection reagent is prepared by dissolving 0·2 g of ninhydrin in 100 ml of acetone. Immediately before use a little 2% pyridine is added. The reagent may then be sprayed onto the chromatogram, or the chromatogram may be dipped into the reagent. In either case, immediately after the application of the reagent, heat the chromatogram in an oven at 103°C for 2-3 minutes. The position of the amino acids will be marked by the appearance of violet coloured spots. If no oven is available the chromatogram should be left at room temperature for 3 hours.

Observe any differences in colour produced by the different amino acids and determine their respective Rf values.

$$Rf = \frac{\text{distance travelled by compound}}{\text{distance travelled by solvent}}$$

Note: The ninhydrin reagent is extremely sensitive and care should be taken in handling the chromatogram as the reagent will detect the small traces of amino acids left on the paper by the fingers. Similarly care should be taken to avoid getting the reagent on the hands as the stains are extremely difficult to remove.

Experiment 56

To demonstrate the separation of sugars by descending paper chromatography

Note: Sugars are extremely slow moving compounds and are best separated by descending paper chromatography (see page 84).

APPARATUS REQUIRED
chromatography tank (see page 85)
Whatman No. 1 grade chromatography paper
samples of:— a. sucrose
                b. glucose
                c. fructose   prepared as 0.7% solutions in 10% isopropanol
                d. arabinose
                e. raffinose
5 capillary pipettes
detection reagent:—p-anisidine phosphate or silver nitrate reagent   5 ml)
separating solvent:—n-butanol: pyridine: water (80 ml: 20 ml: 10 ml) or n-butanol: acetic acid: water (12 ml: 3 ml:

METHOD OF PROCEDURE
The chromatogram is prepared with the starting line some 10 cm from the top of the paper. The bottom edge of the paper is serrated to allow the solvent to pass evenly off the end of the paper, since the sugars would not be separated by the time the solvent reached the end of the paper (see page 86). Mark off the positions at which the sugars are to be applied and label each accordingly. As well as placing pure samples of the sugars on their respective marks place on one mark a mixture of all 5.

When the spots are dry place the chromatogram in the tank as described on page 85, and add the solvent to the trough. Develop for 18 to 24 hours and then remove and dry in a fume cupboard.

DETECTION OF THE SUGARS
Various detection reagents are available but many contain strong acids or have unpleasant odours. One which is reasonably pleasant to use and has the advantage that it does not require heat for colour development is the Silver nitrate reagent, which consists of two solutions applied consecutively and is prepared as follows:—

Solution A is prepared by adding 0.1 ml of a saturated solution of silver nitrate to 20 to 100 ml of acetone. If a precipitate of silver nitrate appears add water dropwise until it redissolves.
Solution B is a 0.5% solution of sodium hydroxide in ethanol and is prepared by dissolving the sodium hydroxide in 15 ml of water.

This aqueous solution is then cooled and when cold, added slowly to 85 ml of ethyl alcohol.

The silver nitrate reagent is applied first and the acetone allowed to evaporate from the paper before the alkaline solution is applied. All sugars will produce dark brown to black spots in less than 10 minutes in the cold. The background colour will gradually change from yellow to brown so that it is best to mark the position of the sugars by ringing them with a pencil.

Note: The silver nitrate reagent is not specific for sugars and can only be used to detect them in pure solutions. It is not suitable for use in detecting sugars present in plant extracts.

An alternative detection reagent which may be used to detect sugars in plant extracts is the reagent p-anisidine phosphate, which has the added advantage that it produces different colours with the various sugars. This reagent however has to be heated to 90°C before it reacts with the sugar and can therefore only be used where there is an oven or a hot plate available for heating the chromatogram.

The p-anisidine phosphate reagent is prepared as follows:—

0.5 g of p-anisidine is dissolved in 2 ml of ortho-phosphoric acid. The mixture may be heated gently to assist the solution of the solid. The mixture is then diluted with ethanol until the total volume is 50 ml. A fine precipitate forms which may be removed by filtration. The filtrate is used as the detection reagent.

To produce the colours, the chromatogram must be heated at 95-100°C for 3-5 minutes. Hexose sugars yield light brown colours, pentoses reddish brown (dark) and fructose and any sugar containing fructose—yellow, while sugar acids appear pink.

Since the solvent front has left the paper, the Rf values of the sugars cannot be determined. In this case however glucose is used as a reference and the rates of flow of the other sugars are expressed relative to this, (Rg).

$$Rg = \frac{\text{distance substance travels from origin}}{\text{distance glucose travels from origin}} \times 100$$

Determine the Rg's of the other sugars. Compare their speed of movement with their molecular size.

Experiment 57

To demonstrate the separation of the photosynthetic pigments by adsorption chromatography, using a column of icing sugar

APPARATUS REQUIRED

a) for the extraction of the pigments
5 g of dried nettle powder
90% acetone
measuring cylinder
filter funnel
filter paper
stand
2 stoppered flasks
100 ml beaker
water bath
petroleum ether
distilled water
anhydrous sodium sulphate

b) for the chromatography
petroleum ether
glass column
rubber bung through which passes a short length of small bore glass tubing
small bore rubber tubing
adjustable clip
glass wool
icing sugar
teat pipette
beaker
glass rod
stand

METHOD OF PROCEDURE

a) Extraction of the pigments

Weigh out about 5 g of dried nettle powder, and extract with 35 ml of 90% acetone for 15 minutes, in a stoppered container. During the extraction shake the flask gently. Filter into a measuring cylinder and note the volume of the filtrate. Put into a separating funnel a volume of petroleum ether equal to **twice** the volume of your extract, and then add the extract. Stopper the flask and shake well.

The petroleum ether will take up most of the dissolved pigments from the acetone. Pour distilled water, **equal in volume to your original extract,** slowly down one side of the funnel. Do not shake but swirl gently. Run out and discard the lower layer (water + acetone). Repeat with 3 more portions of water to remove all the acetone.

Leave the extract to dry over a tablespoon of anhydrous sodium sulphate in a stoppered flask for at least 15 minutes. It is essential that the extract be kept dry otherwise future experiments involving its use will be spoilt. Pour the extract into a 100 ml beaker and evaporate it over a water bath, to a few millilitres.

Note: A number of chemicals used in this experiment are highly inflammable so that the use of naked flames should be kept to the minimum. The water bath should be heated and the bunsen flame extinguished before the petroleum ether is evaporated.

b) Preparation of the column

Place a small plug of glass wool, not too tightly, at the bottom of a glass column which has been fitted with a bung through which passes a short length of small bore glass tubing. To this glass tubing fit a short length of rubber tubing. To make the joint between the glass and rubber tubing secure it may be wired. Attach the adjustable clip to the rubber tubing. Pour a little petroleum ether into the column and adjust the clip on the tubing so that the liquid drips through slowly. Mix some icing sugar, in a small beaker, with some petroleum ether and pour it into the column. Avoid preparing too thick a mixture. As the sugar settles in the column, leaving a clear layer of petroleum ether above it, pipette off some of the liquid, using a teat pipette, and pour in more sugar suspension. Repeat this process, gently tapping the column to help the sugar settle, until it is evenly filled to within a few centimetres of the top, and then close the tap until you are ready to add your pigment extract.

Diagram showing column before addition of extract

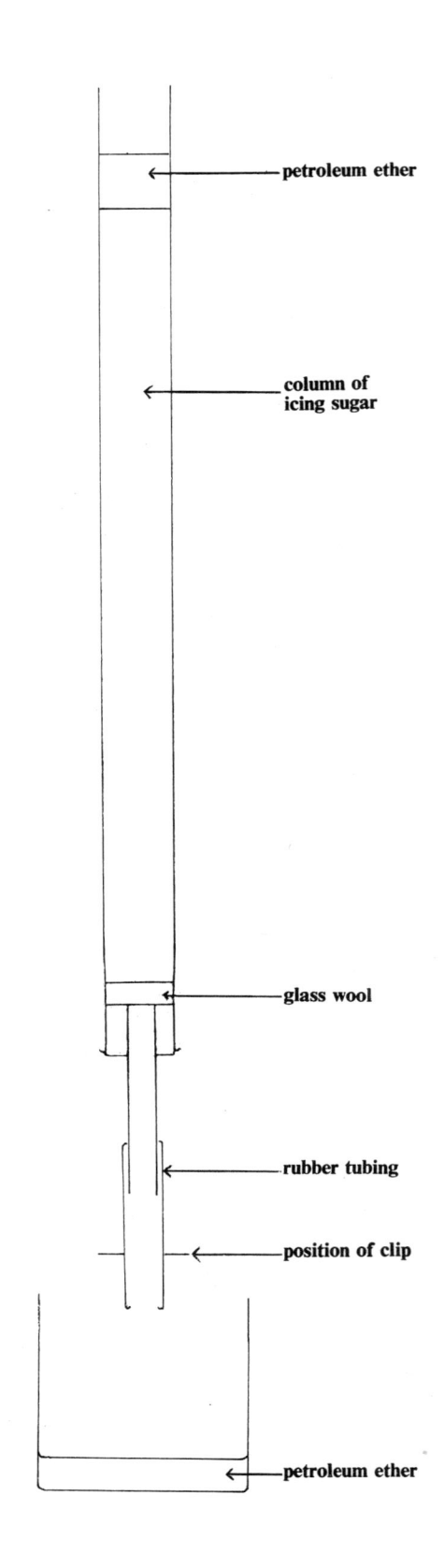

- petroleum ether
- column of icing sugar
- glass wool
- rubber tubing
- position of clip
- petroleum ether

Diagram showing column with separated bands of pigment

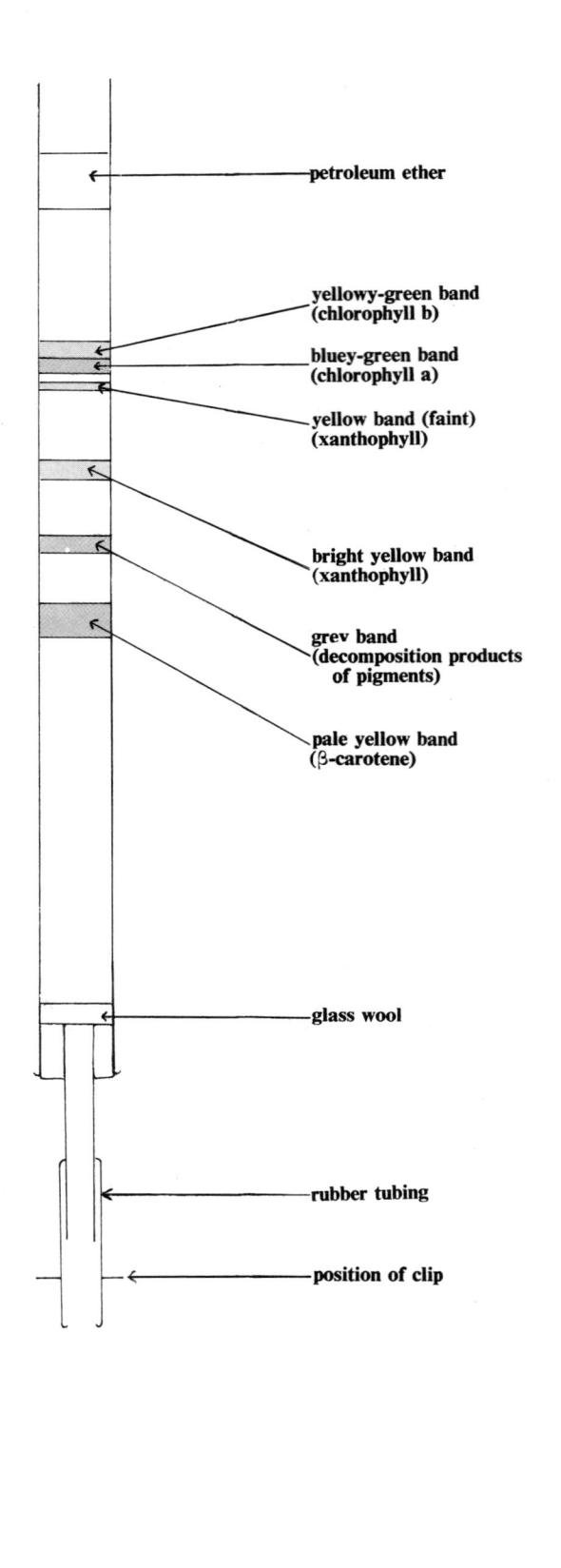

- petroleum ether
- yellowy-green band (chlorophyll b)
- bluey-green band (chlorophyll a)
- yellow band (faint) (xanthophyll)
- bright yellow band (xanthophyll)
- grey band (decomposition products of pigments)
- pale yellow band (β-carotene)
- glass wool
- rubber tubing
- position of clip

c) Chromatography of the extract

When the pigment has been concentrated sufficiently (see (a) above), remove the excess petroleum ether from above the column, using a pipette, then **gently** add the extract to the top of the icing sugar column. Open the clip at the bottom of the column so that liquid again runs through slowly. When nearly all of the extract has moved into the column, **gently** add more petroleum ether to the top of the column. As the extract moves down through the column it separates into distinct bands of pigment. Watch the progress of the bands down the column, adding more petroleum ether as needed to prevent the column running dry. When the pigments have separated sufficiently (bands of carotene, xanthophyll, chlorophyll a and chlorophyll b clearly visible), close the clip and make a drawing of them.

Note: It is essential that during development the column should not be allowed to run dry. Always keep some petroleum ether above the column.

Experiment 58

To demonstrate the separation of the photosynthetic pigments by paper chromatography, using the ascending technique

APPARATUS REQUIRED

a) for the extraction of the pigments
as for experiment 57 on page 91.

b) for chromatography
chromatography tank (or large sweet jar)
Whatman No. 1 grade chromatography paper
capillary pipette
developing solvent:—acetone: petroleum ether (5 ml: 45 ml)
white petroleum jelly
paper clips

METHOD OF PROCEDURE

The pigment extract is prepared as described in experiment 57, and is applied to the chromatogram, which has been prepared in the way described on pages 84 & 87. The chromatogram is then rolled into a cylinder using two pairs of linked paper clips (see page 87), and placed in the chromatography tank with the end carrying the sample spots dipping into the solvent. Make sure the sample spots themselves are not covered by the solvent. Replace the lid sealing it with white petroleum jelly and leave the chromatogram to develop.

Make a diagram showing the position of the separated compounds on the paper and determine their Rf values.

Experiment 59

To demonstrate the separation of the photosynthetic pigments by thin layer chromatography

APPARATUS REQUIRED
grease free microscope slides
cellulose MN300 powder for chromatography
distilled water
petroleum ether
acetone
2 beakers
foil, petri dish or watch glass
extract of photosynthetic pigments (see page 91)
fine capillary pipette

METHOD OF PROCEDURE

Preparation of the thin layer chromatogram

Prepare a thick cream from about 1 g of the cellulose powder and a few drops of distilled water. Carefully pour some of this on to a glass slide so that it forms a pool across the end of the slide (see first diagram page 94). It is essential that the glass slide be free from grease. Take a second glass slide, and, holding it very loosely between forefinger and thumb, (as shown in second diagram), dip the end into the pool of adsorbent on the first slide and draw it **very lightly** along the length of the slide to spread the layer. The best results will be obtained by allowing the "spreading" slide to hinge very slightly between finger and thumb as it is drawn along. The prepared slide must be allowed to dry completely before being used. This takes about 15-20 minutes at room temperature, but better separations will be obtained if the slides are heated in an oven at 100°C for 10 minutes.

Diagram 1

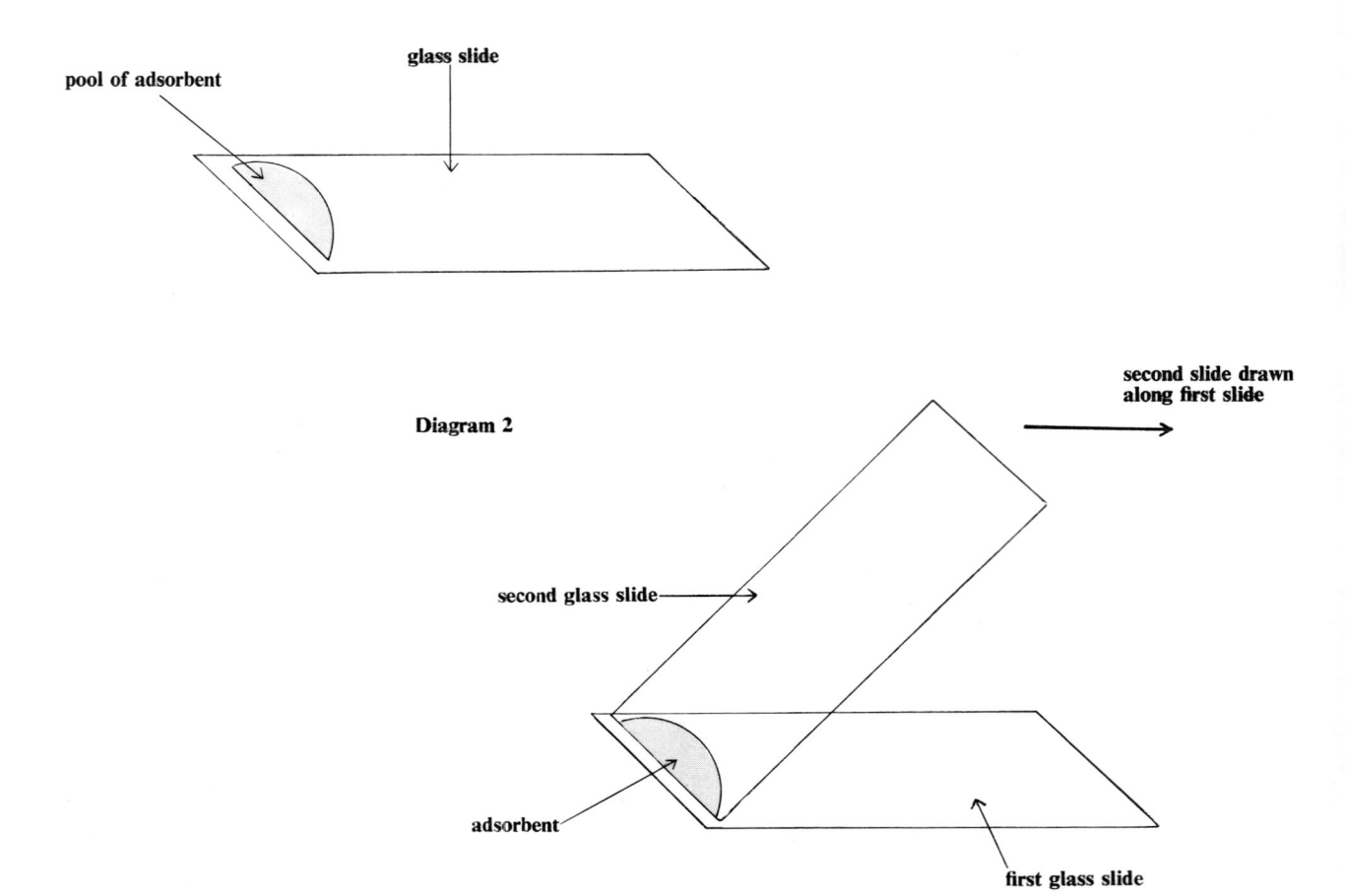

glass slide

pool of adsorbent

Diagram 2

second slide drawn along first slide

second glass slide

adsorbent

first glass slide

An alternative method of preparing the layer is to place a slide between two other slides which have been slightly raised by placing below them a thin piece of plastic or similar material, see diagram below.

Diagram showing second method for preparing the thin layer chromatogram

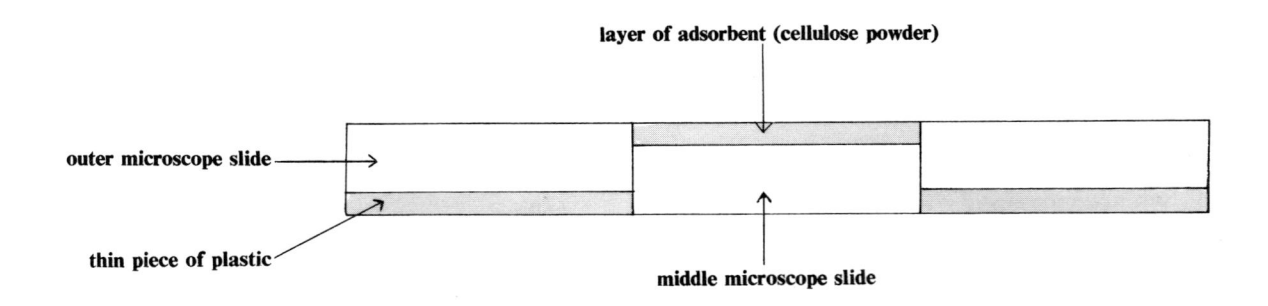

layer of adsorbent (cellulose powder)

outer microscope slide

thin piece of plastic

middle microscope slide

The adsorbent cream is then added to one end of the middle slide and spread by a fourth slide which is drawn over the middle slide while resting on the outer two slides. The thickness of the layer will be the difference between the heights of the outer and middle slides.

After the pigment extract has been applied to the layer using a fine capillary pipette it is left to dry, the plate may be developed by placing it, with the sample spot lowermost, in a beaker containing a few millilitres of a mixture of petroleum ether and acetone (75 parts to 25 parts).

The top of the beaker is covered with foil, or a petri dish, to prevent the escape of solvent vapour.

Diagram showing development of thin layer chromatogram

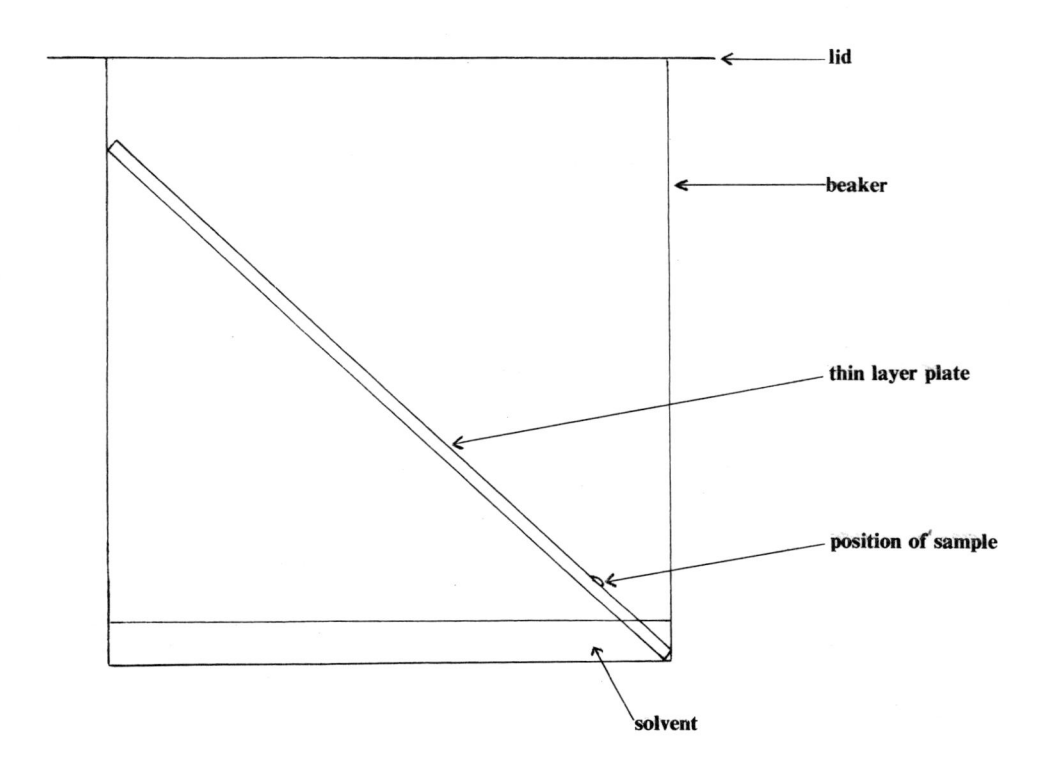

Development takes only a few minutes and it is possible to observe the separation of the pigments as it takes place.

Compare the speed of separation and sensitivity, with the previous method, experiment 58.

Experiment 60

To demonstrate the presence of sugars in plant tissues using descending paper chromatography

APPARATUS REQUIRED
pea seedling roots
electric hot plate
conical flask
2 microscope slides
6 capillary pipettes
chromatography tank
Whatman No. 1 grade chromatography paper
developing solvent:—n-butanol: pyridine: water (8:2:1)
80% ethanol
standard solutions of glucose, fructose, sucrose, xylose and raffinose
p-anisidine phosphate prepared as in experiment 56 on page 90
white petroleum jelly

METHOD OF PROCEDURE

The sugars may be extracted from the plant material either with hot 80% ethanol or by expressing the sap directly onto the chromatogram. (Greenshield's technique)

a) Ethanolic extraction

The plant material is extracted with boiling 80% ethanol for 15 minutes. (In this experiment pea seedling roots are used to avoid having to remove the photosynthetic pigments). The ethanolic extract is then filtered and concentrated on an electric hot plate before being applied to the chromatogram, in the manner previously described on page 85.

b) Greenshield's method

The seedling roots are divided into 2 or 3 sections, each 2-3 mm thick. Place a section at a marked position on the starting line of the paper chromatogram. Place a microscope slide under the paper, and below the section. Place another slide on top of the section and express the sap by exerting pressure on the top slide. When the sap has been absorbed by the paper and has spread to a spot 5-6 mm in diameter, dry it rapidly in a stream of warm air from a hair dryer. Repeat the process with another section of seedling root.

As well as samples of the extract or sap, place at different positions on the starting line 1 drop of each of the following:— sucrose, glucose, fructose, xylose, and raffinose. The bottom edge of the chromatogram is serrated to allow the solvent to pass evenly off the end during development. The chromatogram is then placed in the tank, the solvent added, and the lid replaced and sealed in place with white petroleum jelly. After development (18-24 hours) the chromatogram is removed, dried in a fume cupboard and the location reagent (p-anisidine phosphate, see page 90), applied. The chromatogram is then heated to 90-100°C for 2-5 minutes and the colour and position of the sugars present in the extract compared with those of the marker sugars.

Note: Compounds in plant extracts frequently run a little slower than the same compounds in pure solutions.

Experiment 61

To demonstrate the presence of amino acids in plant extracts using ascending chromatography

APPARATUS REQUIRED
potato tuber
Whatman No. 1 grade chromatography paper
filter funnel and filter paper
mortar and pestle
6 capillary pipettes
test tube
water baths
paper clips
chromatography tank
80% ethanol
developing solvent:—n-butanol: acetic acid: water (30 ml: 7·5 ml: 12·5 ml)
location reagent:— ninhydrin reagent
standard solutions of tryptophan, glycine, leucine, and valine
white petroleum jelly

METHOD OF PROCEDURE

Macerate 2-3 g of potato tissue with 10 ml of 80% ethanol, and leave to stand for 15 minutes. Filter into test tube and concentrate the filtrate over a water bath.

Prepare the chromatography paper in the way previously described on page 87 and apply some of the concentrated filtrate at a marked position on the starting line. Also apply at different positions along the starting line, the pure samples of the amino acids. Roll the chromatogram into a cylinder as described on page 87, using paired paper clips to hold it in position, and place it, with the sample spots lowermost, into the chromatography tank so that its lower edge dips into the solvent, but the sample spots are not immersed in the solvent. Replace the lid of the tank sealing it with white petroleum jelly and develop for at least 4 hours.

Remove the chromatogram from the tank, mark the position of the solvent front, and then dry it in a fume cupboard. When the chromatogram is dry, locate the amino acids with the ninhydrin reagent as previously described on page 89.

Compare the position and colours of the amino acids in the extract with those of the marker solutions. Determine the Rf value of each amino acid detected.

GROWTH

Growth can be defined as an irreversible increase in volume which is usually accompanied by the production of protoplasm, that is, it is an increase in amount of living matter.

As well as increasing in volume during their life plants also change shape, certain parts becoming differentiated and more complex. This change in shape, differentiation and increase in complexity is referred to as the development of the plant. Since growth is an integral part of the overall process of development it is rather difficult to draw a line between growth and development.

Growth occurs at different levels. It occurs at the level of the whole organism, which itself is due to the growth of the various organs. These in turn grow as a result of growth by the tissues of which they are composed and the growth of a tissue is the sum of the growth of its constituent cells. The cell is therefore the ultimate unit of independent growth, since the cell organelles only grow inside the cell as part of cell growth.

There are many ways by which the growth of a plant may be measured. One of these is by determining the increase in the plants' dry weight, since this increase in weight is the result of the synthesis of extra protoplasm. If the increase in dry weight of a plant is followed during its life cycle, it will be found that the rate of increase, and therefore the rate of growth of the plant is not constant, that is the plant does not put on a steady 10 grams in weight each week.

Immediately after germination, the dry weight of the plant decreases slightly, since at this stage, the plant is unable to replace, by photosynthesis, the reserve materials which are being utilised to provide the energy and basic building materials necessary for growth. Once the leaves have been formed however, the growth rate of the plant increases rapidly as photosynthesis becomes established. Growth continues at this rapid rate until the plant nears maturity and then the rate declines and approaches zero. During the final stages of the life of the plant senescence causes a decrease in the dry weight. If this growth pattern is represented graphically it results in an S-shaped or sigmoid curve, (see diagram below), which is typical not only of the growth rate of the whole plant but also of the individual organs. Indeed this type of growth curve is typical of living organisms in general.

Graphical representation of the growth pattern of a plant

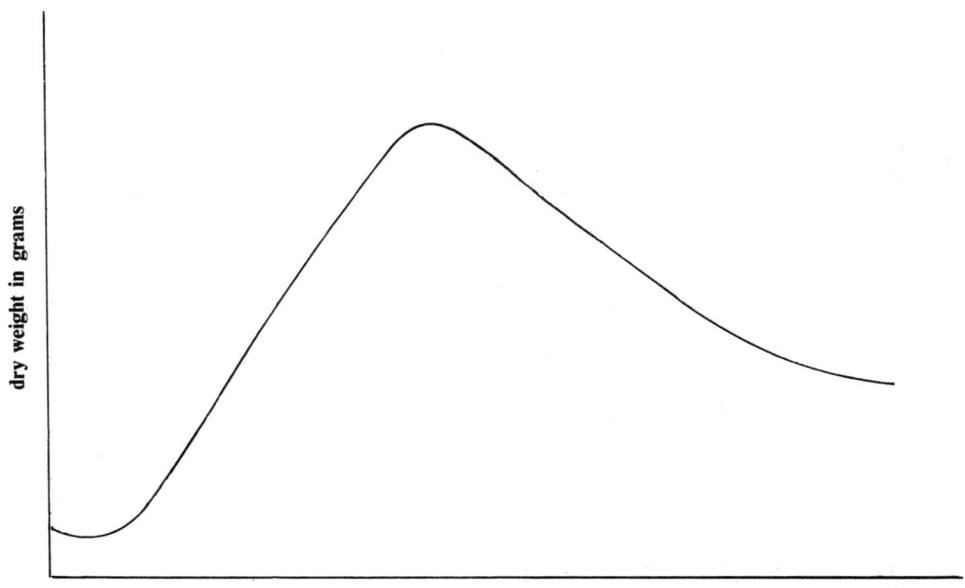

The growth of a plant is the result of new cells being formed by cell division and the subsequent enlargement of these newly formed cells. Unlike animals, plants have their growth confined to limited parts of their bodies. These regions are known as the **meristems** and they are found in various positions in the plant. All plants have a meristem at the extreme apex of each root and shoot and often each leaf. These meristems are termed **apical** or **primary meristems** and the growth resulting from their action, **primary growth.** The activity of these meristems results in an increase in length of the organ. Many plants also possess **secondary,** or **lateral meristems** which serve to increase the diameter of the organ, **secondary growth.** In addition to these types of meristems, some plants, particularly grasses and cereals, also have meristems which are said to be **intercalary** since they occur in the **internodal regions** of the stem. (See also Book 1, page 9).

A longitudinal section of the apical meristem of a shoot shows it to have an apex of small cells. These cells constitute the **promeristem** and by their continued division they produce the cells which ultimately form the various tissues of the shoot. The cells of the promeristem are all similar in appearance and have very thin walls, dense, non-vacuolated cytoplasm and relatively large, prominent nuclei. Immediately below the promeristem is a zone of cells which are larger but still engaged in active division. In this region, lateral protuberances are produced at regular intervals. These are the leaf primordia and will eventually develop into the foliage leaves.

A similar longitudinal section of a root apex shows that it too possesses a promeristem, but that the meristem is in this case protected from injury by a cap of cells, the root cap. This is a necessary precaution since the root grows through soil and not the air as does the shoot. As in the stem the promeristem gives rise to cells which, as they become older, elongate and differentiate and as a result there is an increase in the length of the organ.

In both stem and root this pattern of longitudinal development is largely determined by the fact that when the cells of the promeristem divide, they do so in such a way that the new cell wall is laid down at right angles to the direction of growth, thus cutting off cells to the rear. The process of cell division in meristems **(mitosis)** has been one of the most intensively studied processes in biology since its discovery around 1880, and yet our knowledge of it is still very largely limited to a description of the physical events which take place. We know very little of the actual physiological processes which initiate cell division.

Mitotic cell division
For descriptive purposes, the process of mitosis can be divided into five stages:— Interphase, Prophase, Metaphase, Anaphase and Telophase. This division is of course artificial since the process is a continuous one.

Interphase is the stage between two successive divisions, when the chromosomes appear dispersed and the nucleus is visible as a definite structure surrounded by the nuclear membrane. Because very little appears to be happening this stage is often referred to as the **'resting stage'**, a description which is far from accurate as we shall see later, since physiologically the interphase period is perhaps the most active of all the stages.

As the cell moves from interphase to **prophase,** the nucleus increases in volume and becomes more granular. In the early stages of prophase, the individual chromosomes become visible as thin threads which, as prophase proceeds, become shorter and thicker as a result of becoming coiled or "spiralised". At this stage it is possible to observe that each chromosome in fact consists of two threads (the **chromatids**) which are joined at a single point, the **centromere.** Some sort of attractive force seems to exist between the two chromatids of a single chromosome (the sister chromatids) since they remain very closely paired along their entire length, and do not drift apart. The end of prophase is marked by the gradual disappearance of the nucleolus and the sudden breakdown of the nuclear membrane.

A characteristic feature of **metaphase** which follows prophase is the development in the cytoplasm of a series of fibrous threads. These fibres, which are proteinaceous, lie roughly parallel to each other and extend from one end of the cell to the other. Because of the general shape and appearance of this structure it is known as the **spindle.** During the formation of the spindle the chromosomes become even thicker and shorter and move through the cytoplasm until they become arranged in a central mass on the equator of the spindle. Here they become attached by their centromeres to the spindle fibres. It is noticeable that at this stage the attractive forces existing between the sister chromatids seems to have lapsed since the chromatids are no longer closely paired.

Anaphase begins with the division of the centromeres so that the sister chromatids are no longer attached to each other. The sister chromatids, or daughter chromosomes as they are now called, move slowly towards opposite poles of the spindle. The separation seems to be due partly to the repulsive forces which develop between the sister chromatids, partly to the contraction of the spindle fibres towards the poles and partly to the expansion of the central region of the spindle itself.

Telophase begins when the two groups of chromosomes have reached the poles and the events which occur are essentially the reverse of those which occurred during prophase. The chromosomes despiralise and revert to their interphase condition. Each group becomes surrounded by a nuclear membrane which is resynthesised from fragments of the endoplasmic reticulum, and the nucleoli are reformed.

Now that nuclear division is completed, **cytoplasmic division** begins. The first indication that the cytoplasm is about to divide is the appearance of a structure called the **phragmoplast,** in the equatorial region of the spindle. The **cell plate** is formed in the central parts of the phragmoplast and grows towards the periphery of the cell until it reaches the cell walls. Further development involves the laying down of a **cell wall** along the cell plate, which therefore becomes the **middle lamella.**

Since the new cell wall is always laid down at right angles to the long axis of the spindle, it follows that the orientation of the spindle has a profound influence upon the direction in which the daughter cell is cut off.

At the end of telophase, the daughter cell nuclei have reverted to the interphase state. It is interesting to note that at the end of telophase, the chromosomes are single stranded structures and yet at the beginning of prophase, they appear double stranded. The extra material must therefore be synthesised during the interphase period. It has also been shown that the material from which the spindle is constructed is also synthesised during interphase. In view of this high metabolic activity it would seem to be more suitable to term interphase the **"metabolic"** rather than the "resting" stage.

Immediately behind the meristematic region of the stem or root is the region of elongation where the extension of the cells is the dominant process. Although the amount of extension varies from plant to plant, it is not uncommon for a cell to show a 30-40 fold increase in its length. This increase in length is a fairly rapid process, the individual cells completing their expansion within three to four days. The rapid increase in length by the cells leads to the equally rapid growth of the organ. For example it has been estimated that the root of the broad bean plant grows at the rate of 2·2 cm a day but even this is slow when compared with the growth of bamboo shoots which average 58 cm a day.

The precise mechanism behind cell extension has not been fully elucidated, but it has been suggested that a rapid uptake of water by expanding cells leads to an increase in the volume of the cells, and as a result the cell walls become stretched. This does not itself constitute growth since the stretching is reversible and there has been no increase in protoplasm. Growth occurs when the cell wall is stretched beyond its elastic limit and new cell wall material is incorporated, thus making the extension permanent. Another consequence of the uptake of water is the vacuolation of the cell, and limiting of the cytoplasm to a thin peripheral layer lining the cell wall.

Behind the region of elongation lies the region in which the elongated cells become differentiated, that is, where they become modified structurally so that they can perform specific tasks within the plant. It is not clear how the process of differentiation is controlled, but research workers investigating the number and nature of the enzymes present in cells made the interesting discovery that each type of cell had a specific enzyme complement and that this complement differed from that found in the other types of cell. In some cases the difference was due to one type of cell possessing different enzymes, while in other cases, the same enzymes were present but in differing amounts. Cell type A, for example, might have the enzymes X, Y and Z present with enzyme X at a very high concentration and the other two enzymes at only low concentrations. Cell type B also has the enzymes X, Y and Z, but in this case it is the enzyme Y which is present at a high concentration and enzymes X and Z at low concentrations. In cell type C, however, enzyme Z is the most abundant enzyme with X and Y present only sparingly. Since the overall pattern of a cell's enzymic activity depends on the balance between the various enzymes present, these three types of cell would have different patterns of enzyme activity (enzymic patterns) even though they possess the same enzymes. Since it has been shown that cells differentiating into phloem tissue possess different enzymic patterns from those differentiating into xylem tissue, it has been suggested that differentiation is the result of the controlled production of these enzymic patterns. How exactly this is done is a problem which still awaits elucidation.

By now it should be clear that our knowledge of the processes of growth and differentiation at the cellular level is, at best, very rudimentary. We know the overall facts but not the fine details. What is also obvious is the fact that the processes are under a definite control. This is also true at the tissue and organ levels of growth. The development of the plant is the result of the growth of the various organs, being controlled and integrated in such a way that a plant of definite form and size results. How is this integration of growth achieved? There is a great deal of evidence supporting the view that the integration of the activities in one part of the plant, with the activities in other parts, is the result of the synthesis, transport and utilisation of specific chemical compounds which have been termed **plant growth hormones.**

With very few exceptions these growth compounds are produced in the apical meristems of the plant, which therefore play a major role in determining the pattern of growth and development of the plant. This has been amply demonstrated in experiments in which root tips were excised and replaced after re-orientation. The new tissues as they developed were not in line with the tissues of the root stump. Therefore the pattern of tissue development is laid down in the tip.

The first person to demonstrate the importance of the apical meristem to the growth of the plant was Charles Darwin, whose actual interest lay in the response of plants to light. He had observed that plants illuminated from one side only, always grew in such a way that they eventually faced towards the light. He chose for his work the coleoptile or leaf sheath of the oat seedling and found that when he shaded the tip of the seedling but illuminated the rest, the coleoptile did not respond to light. When however he illuminated only the tip, the coleoptile grew towards the light. What Darwin found curious was the fact that the region which received the light, that is the tip, was able to control the reaction of cells in the growing region, which in oat seedlings is some distance away from the tip.

Some 40 years later, in 1919, a Hungarian biologist by the name of Paal suggested, with remarkable foresight, that the tip of the root or shoot produced a substance or substances which pass back through the living tissues behind the tip and control the growth of cells there. Under normal circumstances the substances passing back are equally distributed on all sides of the organ so that growth is also equal on all sides. However, should the movement of these growth controlling substances be disturbed on one side of the organ then the growth on this side decreases and the organ becomes curved.

The experiments of Söding in 1923, in which he removed the tip of the coleoptile, served to strengthen this hypothesis. The decapitated coleoptiles stopped growing and only resumed elongating when the excised tips were replaced.

The elusive growth substances were first extracted from the coleoptiles by Went (1928), who used the simple technique of placing the excised coleoptile tips on small blocks of agar and allowing the substances produced by the tips to diffuse into the agar. When these agar blocks were placed on the decapitated coleoptile stumps, the coleoptiles resumed growth, thus showing that the growth substances had in fact passed into the agar. By using agar blocks prepared in this way, Went was also able to show that an unequal distribution of the auxin within the coleoptile would produce unequal growth and curvature. When he placed an agar block symmetrically on the coleoptile stump, the stump grew vertically. When he placed the block eccentrically, so that only one side of the coleoptile received the growth substances, the cells on this side elongated more rapidly than the cells on the other side thus producing a curvature of the coleoptile.

Diagrams showing curvatures resulting from the placing of agar blocks containing growth hormones on decapitated oat coleoptiles

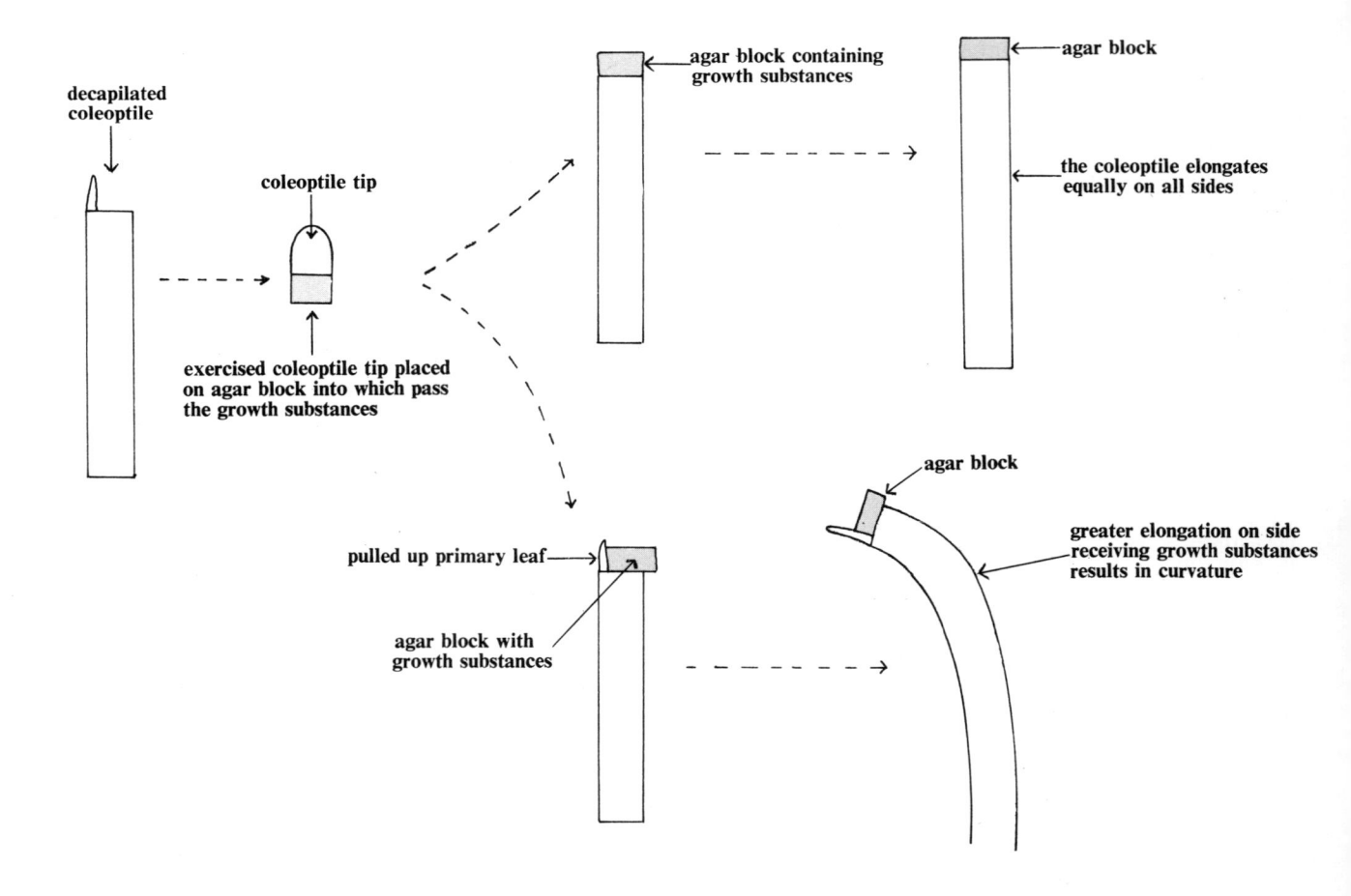

Once the existence of a growth substance had been unequivocally demonstrated, the elucidation of its chemical nature was the next logical step. So far it has proved impossible to identify the actual compound present in the coleoptile tip, chiefly because the amounts present are very minute. However, a compound has been isolated from other plant tissues which is extremely active in producing curvature in the oat coleoptile. Furthermore it has also been shown that the coleoptile tip possesses all the enzymes necessary to synthesise the compound and it is now generally accepted that the primary growth controlling substance of the coleoptile, and indeed of all higher plants, is this compound, β-**indolyl acetic acid** (IAA); which is generally referred to as **heteroauxin.**

β-indolyl acetic acid (IAA)

—CH$_2$ COOH

N
H

Although IAA is the primary growth controlling compound present in higher plants it is by no means the only one. Chromatography has revealed the existence of a number of other "auxins" all of which are indole derivatives. It is possible that many of these compounds are in fact intermediates of IAA synthesis, but the picture at the moment is more than a little confused.

Evidence has continued to accumulate that auxin is not confined in its physiological activities to the control of cell extension. It can, for example, promote the formation of adventitious rootlets at the base of stem cuttings, and has also been implicated in the induction of fruit formation, and in the controlling of leaf and fruit fall. It is also involved in the inhibition of the growth of lateral shoots and roots, a phenomenon known as apical dominance, and has been found able to promote or inhibit cell division.

In all its activities auxin has been found to be able to promote or inhibit. The exact effect seems to depend on the concentration of the auxin and the sensitivity of the tissue. Shoots for example are far less sensitive to auxin than are roots. Concentrations of auxin which promote the extension of shoots are highly inhibitory to root extension, as can be seen from the diagram below.

Graphical representation of the effect of auxin concentration on the growth of roots and shoots

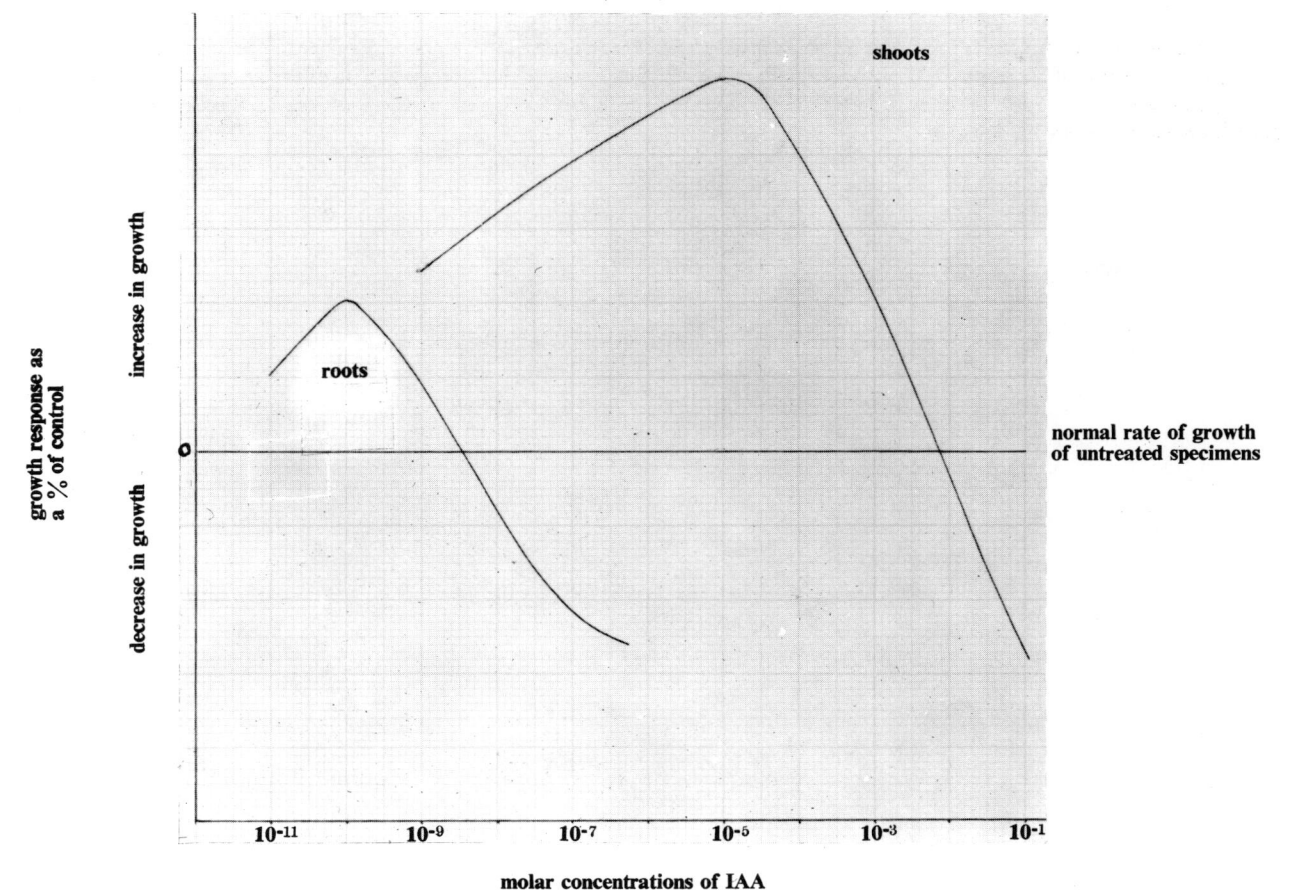

For a long time the auxins remained the only naturally occurring growth substances known to exert an influence on the development of plants. Then in 1926, the attention of research workers in Japan was drawn to rice plants infected by the fungus Gibberella fujikuroi. These plants were considerably taller than the uninfected seedlings and, what was more important, an extract of an infected plant was found to be able to stimulate the growth of other species of plants. In 1939, Yabata, Hayashi, and Sumiki succeeded in isolating the active constituent of the extract as a crystalline compound which they called **Gibberellin A**. Later work revealed the presence in the extract of five closely related substances, which are known as **Gibberellins A_1, A_2, A_3 (Giberellic acid), A_4 and A_5**, and subsequent research indicates that many of these gibberellins and gibberellin like compounds are of widespread occurrence in higher plants.

Gibberellic acid (A_3)

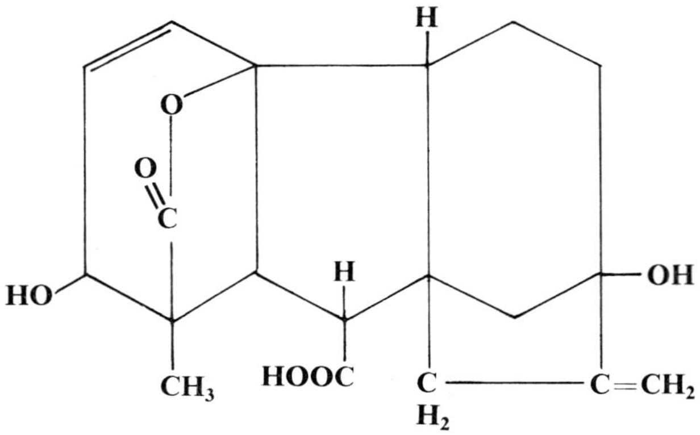

The chief biological property of gibberellins is their ability to increase the length of the cells and therefore the length of the internodes, (particularly those of dwarf strains) but they have also been shown to be active in initiating cell division. This has led some workers to suggest that the gibberellins initiate division in the cambial cells and auxin controls the differentiation of the cells produced.

Shortly after the discovery of the gibberellins came the discovery by Skoog of yet another class of growth regulating compounds. Since these compounds were very active in the promotion of cell division they were called **kinins**. The kinin originally isolated by Skoog was a breakdown product of nucleic acids and was called **6-furfurylaminopurine** or **kinetin**.

Kinetin

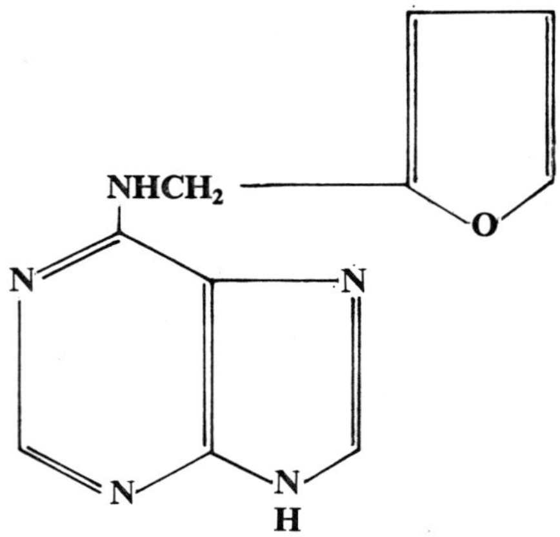

Although no naturally occurring kinin has yet been isolated from plants, it has been demonstrated on several occasions that the cells of tissue cultures will not develop in a proper manner unless auxins, gibberellins and kinins are present. Furthermore it has been shown that the relative concentrations of these compounds determines whether the callus tissues develop root or shoot meristems. Since this has been demonstrated in tissue cultures it has been suggested that a similar balance exists in the plants themselves.

A large number of synthetically prepared compounds have been shown to possess activity in the regulation of growth, indeed some of these are even more active than the naturally occurring compounds. These **"synthetic auxins"** have been used extensively in the rooting of cuttings and as selective herbicides. One reason for this lies in the fact that the plants do not possess enzymes capable of breaking down these compounds so that relatively high concentrations can be used. If high concentrations of IAA were to be used the plant would merely break down the excess auxin. One of the most widely used **"herbicidal auxins"** is 2.4.dichlorophenoxyacetic acid (2.4.D).

2.4.dichlorophenoxyacetic acid

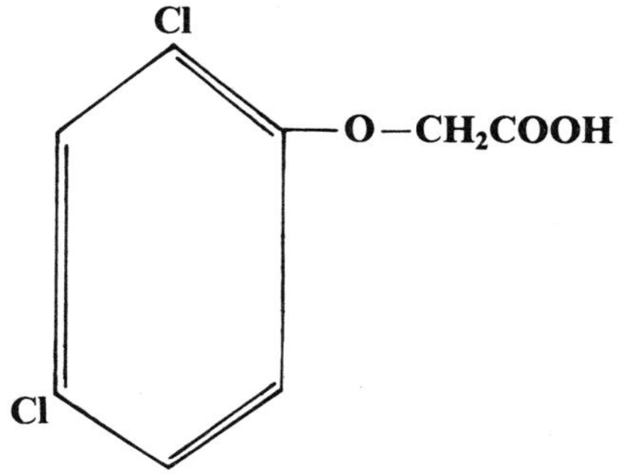

This compound owes its extensive use to the fact that it affects dicotyledonous plants to a far greater extent than it does monocotyledonous plants. Concentrations which are toxic to dicotyledonous plants, such as charlock, do not harm monocotyledonous crop plants such as wheat, maize, oats, etc. Furthermore 2.4.D decomposes after a very short time in the soil and does not have any toxic effects on animals, unlike some of the other chemical herbicides.

GROWTH EXPERIMENTS

Experiment 97

To determine whether water is necessary for germination

APPARATUS REQUIRED
2 trays or containers
vermiculite or bulb fibre
dry pea seeds
pea seeds which have been soaked overnight, in water

METHOD OF PROCEDURE
Place a few dry pea seeds in a tray containing some dry vermiculite (or bulb fibre). Do not water this tray during the experimental period. Into another tray place some damp vermiculite (or bulb fibre) and some seeds which have been soaked overnight in water. This tray may be watered if the vermiculite dries out. After several days examine both trays and compare the germination of the seeds.

Experiment 98
To determine whether air is necessary for germination

APPARATUS REQUIRED
1 stoppered bottle (preferably screw top)
1 unstoppered bottle
pea seeds which have been soaked overnight, in water
freshly prepared cold, boiled water

METHOD OF PROCEDURE
Completely fill the stoppered bottle with cold, boiled water and drop into it 10 soaked pea seeds. Replace the stopper making sure that it fits tightly. The boiled water contains very little oxygen so that the seeds are enclosed in an atmosphere that is virtually anaerobic. In the other bottle place a few drops of unboiled water and 10 soaked seeds. Keep both bottles in a warm place and examine after several days.

Experiment 99
To examine the effect of temperature on germination

APPARATUS REQUIRED
3 trays, preferably metal
vermiculite or bulb fibre
pea seeds which have been soaked in water overnight

METHOD OF PROCEDURE
Place some damp vermiculite or fibre in the trays and plant 10 soaked peas in each tray. Place one tray in a refrigerator, or cold store room, leave the second tray in a warm room, and place the third tray in an oven at 60°C. Leave all three trays for a few days and then examine the percentage germination in each. Make sure that the vermiculite in the trays does not become too dry during the experiment. In the case of the third tray this can be achieved by covering the tray with some metal foil.

Experiment 100
To determine by direct observation the growth rate of a root

APPARATUS REQUIRED
small specimen tube
germinating pea or bean seed with a straight root
rubber bands or transparent adhesive tape
microscope
micrometer eye piece and calibrated slide

METHOD OF PROCEDURE

a) Calibration of the eyepiece micrometer

An eyepiece micrometer (or graticule) consists of a glass plate having engraved upon it a scale of arbitrary length. This plate is inserted into the eyepiece of a microscope so that the scale lies in the plane of the final image. The alignment of this scale is adjusted by rotating the eyepiece while looking down into the microscope.

A calibrated slide (stage micrometer) consists of a microscope slide having a line of indicated length engraved upon the slide and protected by a coverslip. The length of this line varies but it is usually 100μ long and is divided into subdivisions of 2μ. The stage micrometer can be manipulated like any microscope slide.

Place the eyepiece micrometer in the eyepiece and the calibrated slide on the stage of the microscope. Focus carefully and adjust the scales until they are aligned so that the divisions of one are alongside the divisions of the other. Determine the number of divisions on the eyepiece scale which correspond to a certain number of divisions on the stage micrometer scale. Calculate the absolute value of one division of the eyepiece scale. With the stage micrometer removed the eyepiece scale can be used to measure distances under the conditions which were used for calibration.

$$40 \text{ eyepiece divisions} \equiv 100 \text{ stage micrometer divisions}$$
$$\therefore 40 \text{ eyepiece divisions} \equiv 100\mu$$
$$1 \text{ eyepiece division} \equiv \frac{100}{40} = 2\cdot5\mu$$
$$1 \text{ eyepiece division} = 2\cdot5\mu$$

b) Measurement of growth

Take care to avoid laying the root horizontally or rubbing it before the experiment is set up, otherwise a geotropic or mechanical stimulus is imparted and curved growth occurs. Also take care to prevent the root from drying out.

Place a little water at the bottom of a small specimen tube and balance a seed with a straight root on the rim so that the root hangs freely without touching the sides of the tube. Remove the stage micrometer and turn the microscope horizontally. Place the tube and seed on the now vertical stage and adjust its position until the tip of the root is visible through the eyepiece. Fix the tube firmly to the stage using transparent adhesive tape or rubber bands. Observe and measure the growth of the root using the micrometer eyepiece. Convert your readings to actual distances and plot these on a graph. What is the growth rate in mm/hour?

Experiment 101

To determine the growth rate of a root using a Nielson Jones auxonometer

APPARATUS REQUIRED
Nielson Jones auxonometer
bean seedling with a straight radicle
2 pins
cotton
water

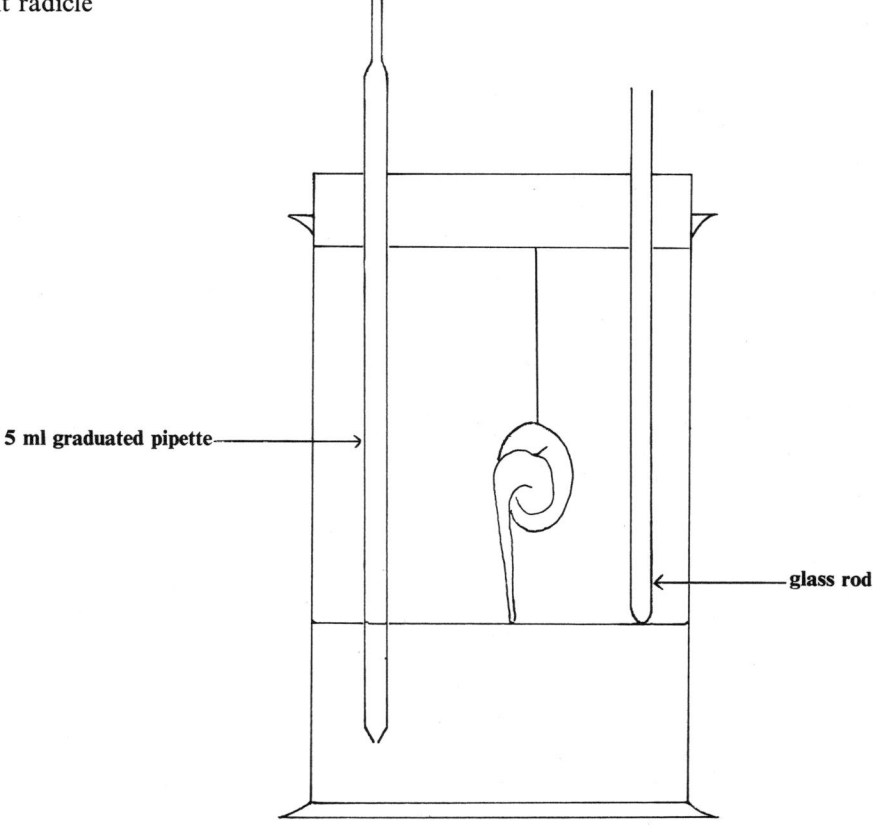

METHOD OF PROCEDURE

An auxonometer is a piece of apparatus which magnifies and records the growth of a plant. The Neilson Jones auxonometer consists of a gas jar fitted with a large cork, through which passes a 5 ml graduated pipette and a long glass rod.

Fill the gas jar about one third full with water. Impale a broad bean seedling on a pin and suspend the bean from the undersurface of the cork, using another pin and a length of cotton, so that the tip of the root just touches the surface of the water. Adjust the glass rod so that this also just touches the surface of the water, as in the diagram.

At 20 minute intervals suck water up into the pipette so that both the root tip and the end of the rod are above the water level. Allow the water to run back slowly into the jar until the water level again just reaches the root tip. Note the volume of the water left in the pipette. Allow more water to run back into the jar until the level now reaches the tip of the glass rod. Again note the volume of water left in the pipette. The difference between the two volumes in the pipette is proportional to the distance between the root tip and the tip of the glass rod. The actual distance between the two tips can be calculated in the following way.

SPECIMEN RESULTS TO SHOW METHOD OF CALCULATING GROWTH OF ROOT TIP

Difference in the two volumes of water in the pipette = 2·97 ml

Cross sectional area of the gas jar = 27 cm^2

$$\therefore \text{distance between the root tip and the glass rod} = \frac{2\cdot 97}{27}$$
$$= 0\cdot 11 \text{ cm}$$
$$= 1\cdot 1 \text{ mm}$$

Since the two tips were originally level, this means that the root has grown 1·1 mm since the beginning of the experiment.

Continue to take readings for 2–3 hours. Does the root grow at a constant rate?

Experiment 102

To demonstrate the zones of growth in a root

APPARATUS REQUIRED
4 bean seedlings with straight radicles
3 glass containers fitted with corks
pins
cotton
Indian ink
ruler
lactophenol containing a **little** cotton blue
microscope slides
microscopes
micrometer eyepiece and stage micrometer
Formalin: Aceto: Alcohol fixative (F.A.A.)
wooden retort stand

METHOD OF PROCEDURE
Select 4 beans with radicles of approximately equal length and measure the length of the radicles. Pin the beans to a stand so that their radicles hang freely. With a thread dipped in Indian ink, mark the radicles at 2 mm intervals. Remove 3 of the beans and hang each, using a piece of cotton and a pin, from the underside of a cork in such a way that the radicles will hang vertically downwards. Place some water in the bottom of the containers and then fit the corks with beans into them. Make sure that the radicles do not dip into the water.

Remove the radicle from the fourth bean. Fix the radicle with FAA and then clear in the lactophenol containing a little cotton blue. If the cotton blue is too concentrated it will make observation of the cells more difficult. The lactophenol may be warmed a little to facilitate clearing of the tissue. When the root has been cleared sufficiently, mount it on a slide and observe under the microscope. Make drawings to scale, using a micrometer eyepiece, of cells from a chosen tissue for example cortex or epidermis, in successive segments.

After 2 days measure the radicles of the first 3 seeds and make scale drawings to see in which zones growth has occurred. Fix and clear one of the roots, and again make scale drawings of cells in the various segments.

By comparing these drawings with the drawings from the first seedling, you will be able to see the changes in cellular dimensions and form during root growth.

Experiment 103

To show the region of elongation in a shoot

APPARATUS REQUIRED
broad bean seedlings (recently germinated)
thread
Indian ink
vermiculite or bulb fibre
pot or tray

METHOD OF PROCEDURE
Germinate some bean seeds in a pot or tray containing some damp vermiculite or bulb fibre. When the seedlings have shoots of only one internode, mark the shoots at 1 mm intervals using a thread soaked in Indian ink. Examine after 2-3 days and determine whether the main region of elongation is at the upper end of the internode or the lower.

Experiment 104

To measure the rate of growth of a shoot using a smoked drum auxonometer

APPARATUS REQUIRED
smoked drum auxonometer
growing bean plant
cotton
cotton wool
xylene
paper
transparent adhesive tape

METHOD OF PROCEDURE
The drum of the auxonometer is covered with a piece of smooth paper held firmly in place with transparent adhesive tape. The smoking of the drum is carried out in a fume cupboard, the drum being suspended over some burning cotton wool which has been soaked in xylene. By carefully turning the drum, it is possible to get a thin layer of soot deposited evenly over the paper.

The apparatus is then set up as shown in the diagram below.

the pull on the top of the shoot should not be excessive

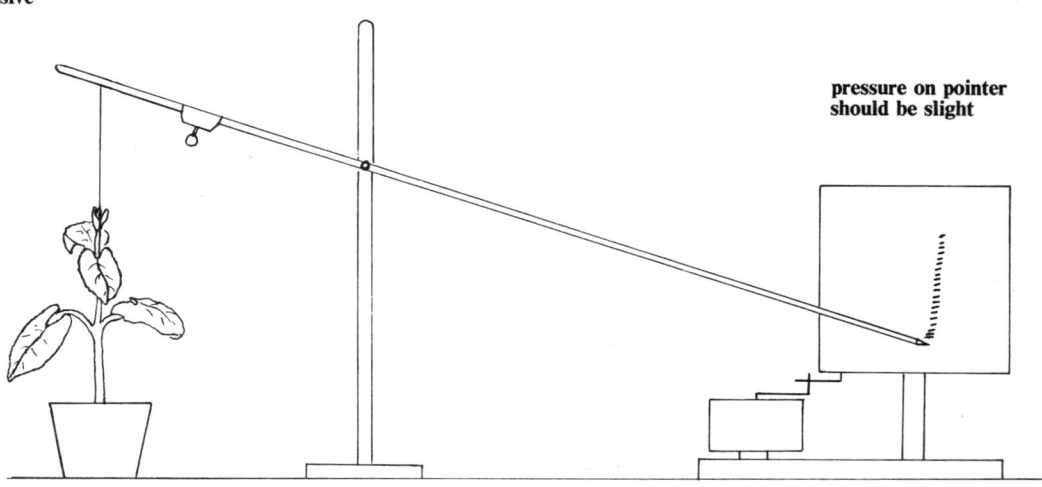

pressure on pointer should be slight

The end of the shorter limb of the pointer is attached to the upper region of the stem of the bean, and the end of the longer limb of the pointer is brought into contact with the smoked drum. As the shoot grows, the end of the pointer in contact with the drum falls and therefore scratches the smoked surface.

Adjust the clock of the auxonometer so that the drum is rotated through a small arc and then returned to its normal position, once every 30 minutes. This results in horizontal marks and allows an accurate record to be made of the rate of growth. The distance between the horizontal marks does not of course represent the true amount of growth, since the long arm of the pointer has been used deliberately to exaggerate this. If the pointer is pivoted so that the length of the pointer on the drum side of the pivot is 10 times that on the plant side, then the distance between the horizontal scratches will be 10 times the actual growth of the plant in that period.

Make a record of the hourly growth rate of the plant over several days and compare the rate of growth during the day with that during the night.

Experiment 105
To measure the growth of a leaf

APPARATUS REQUIRED
actively growing plant in pot, for example a Pelargonium plant
millimetre graph paper
soft pencil
Indian ink

METHOD OF PROCEDURE
Hold the graph paper behind a young leaf of the actively growing plant and trace the outline of the leaf onto the paper, using a soft pencil. Mark the leaf with a dot of ink so that it can be identified easily in future. Determine the area of the leaf by counting the numbers of millimetre squares within the traced outline.

Repeat the process several days running and note the increase in area of the leaf.

Experiment 106
To demonstrate the effect of indolyle acetic acid (IAA) on the growth of oat coleoptiles

APPARATUS REQUIRED
Avena coleoptiles
indolyle acetic acid
phosphate buffer (pH 5·0)
sucrose
distilled water
razor blade
petri dishes
graph paper

METHOD OF PROCEDURE
Germinate some Avena seeds in the dark. At the end of the second day of germination place the seedlings in dim red light (if possible) for a further 24 hours. The coleoptiles are now ready for the experiment. Select coleoptiles which are 2·0, ± 0·2 cm in length, detach them from the seedlings as near to the seeds as possible. Place the coleoptiles on a tile and cut into sections 10 mm long starting at a distance of 3 mm from the tip.

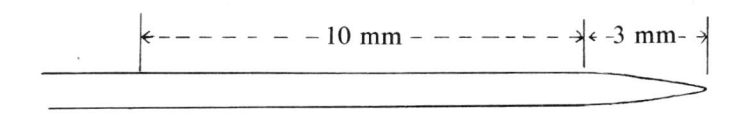

Immediately transfer the sections to a beaker of distilled water. Take great care to avoid squeezing the sections during handling. When 50 sections have been cut they should be thoroughly mixed and then 10 sections should be transferred to each of the 5 test solutions contained in petri dishes.

The test solutions are as follows:—
1) phosphate buffer (pH 5·0) + 2% sucrose.
2) phosphate buffer (pH 5·0) + 2% sucrose + IAA 0·01 mg/l
3) phosphate buffer (pH 5·0) + 2% sucrose + IAA 0·1 mg/l
4) phosphate buffer (pH 5·0) + 2% sucrose + IAA 1·0 mg/l
5) phosphate buffer (pH 5·0) + 2% sucrose + IAA 10·0 mg/l

Place the petri dishes and sections in the dark at 27°C for 24 hours and then remove the sections from the dishes and place them on blotting paper to remove excess solution. Measure the sections against graph paper under a dissecting microscope, and present the results graphically, plotting the increase in coleoptile length against the auxin concentration (log. scale).

Experiment 107

To demonstrate the effect of IAA on the growth of cress roots

APPARATUS REQUIRED
cress seedlings (approximately 30 hours old)
IAA solution (1 mg/litre)
6 petri dishes
filter papers
forceps
graph paper
blotting paper

METHOD OF PROCEDURE
Germinate the cress seedlings by placing them on damp blotting or filter paper in surroundings at 25°C. When the seedlings are approximately 30 hours old, they are ready for use.
From the stock IAA solution prepare 10 ml of each of the following dilutions:—

1) 1·0 mg/l IAA
2) 0·1 mg/l IAA
3) 0·01 mg/l IAA
4) 0·001 mg/l IAA
5) 0·0001 mg/l IAA
6) 0·00001 mg/l IAA

Place a filter paper in a petri dish and add 2·5 ml of solution 1. Repeat with solutions 2, 3, 4, 5, and 6. Select seedlings which have radicles of similar length ($\pm$ 1 mm) and which are preferably 3—5 mm in length, and place 15 seedlings 1 cm from the edge of each dish, with their roots pointing towards the centre. Label dishes clearly and place them in the dark for 24 hours. Then measure the radicles using a piece of graph paper and determine the effect of IAA on the mean extension growth of the roots. Plot your results graphically.

Note: The cress seedling roots are extremely susceptible to damage and should not be handled. The seeds should be transferred by gripping lightly with forceps.

Experiment 108

To determine the effect of IAA on the rooting of plant cuttings

APPARATUS REQUIRED
distilled water
runner bean seedlings
IAA solution (1 mg/l)
polythene sheeting
moist potting compost
trays

METHOD OF PROCEDURE
Take cuttings from runner bean seedlings, divide them into 4 sets and then immediately place the cut ends in the following solutions:—

Set 1   Distilled water
Set 2   1·0 mg/l IAA
Set 3   0·1 mg/l IAA
Set 4   0·01 mg/l IAA

After 3 hours, remove the cuttings, and wash the cut ends in running tap water. Pot the cuttings in moist compost and place under polythene sheeting to prevent wilting. Keep in a warm position, in bright light.
After 7 days carefully remove the plants from their pots or trays and wash off the soil.
Record the number of roots produced by the plants in each treatment.

Experiment 109
To show the effect of gibberellic acid on dwarf pea plants

APPARATUS REQUIRED
10 day old meteor dwarf pea seedlings
micrometer syringe (Agla)
gibberellic acid
80% ethanol

METHOD OF PROCEDURE
Using the 80% ethanol prepare a series of solutions of gibberellic acid so that 0·01 ml of each solution will contain 0·02 mg, 0·08 mg, 0·32 mg and 1·28 mg of gibberellic acid respectively.

Select 10 uniform plants and arrange them into 5 groups. Using an Agla micrometer syringe place 0·01 ml of the appropriate gibberellic acid solution on the upper surface of the first expanded leaf of each seedling in each of 4 of the groups. As a control, place 0·01 ml of 80% ethanol on the leaves of the fifth group of plants. Allow the solution to dry before transferring the plants to a warm position in good light.

Record the height of the plant, number of internodes and the appearance of the plants after 3, 6 and 10 days.

Experiment 110
To demonstrate the phytotoxic effects of 2.4.dichlorophenoxyacetic acid

APPARATUS REQUIRED
5 broad bean seedling plants
2.4.dichlorophenoxyacetic acid (2.4.D)
labels
1·0 ml pipettes
95% ethanol

METHOD OF PROCEDURE
Using 95% ethanol, prepare a series of solutions of 2.4.dichlorophenoxyacetic acid so that 0·1 ml of each solution will contain, 0, 1·0, 10·0, 100·0 and 1000·0 mg of the acid respectively. Apply 0·1 ml of each solution to the youngest fully expanded leaf of the appropriate plant and then label the plants accordingly. Place the plants in a greenhouse or in a warm well lit position, and inspect daily noting any growth effects, particularly their location on the plant.

Experiment 111
To demonstrate the selective phytotoxic properties of 2.4.dichlorophenoxyacetic acid

APPARATUS REQUIRED
4 pots containing wheat seedlings
4 pots having one bean seedling growing in each
solution of 2.4.dichlorophenoxyacetic acid (1000 p.p.m.)
spray
distilled water

METHOD OF PROCEDURE
Take 2 of the pots containing the wheat seedlings and 2 of those containing the bean seedlings and spray all 4 with the solution of 2.4.dichlorophenoxyacetic acid. Take the other pots and spray these with distilled water. Place the plants in a warm well lit position and examine after 1 week.

PLANT MOVEMENT

Since plants are generally sedentary, that is fixed in one place, one tends to overlook the fact that they are capable of a certain amount of movement.

Plant movements may be divided into two types, **autonomic movements** and **paratonic movements.**

Autonomic movements
These are apparently spontaneous and continue for as long as life continues. Examples of this type are the movements of chromosomes during nuclear division; cyclosis, which is the name given to the movement or streaming motion of the cell contents within the cell—this movement is often quite rapid; nutation or, as it is sometimes called, circumnutation, this is the name given to the helical course followed by the apex of the plant as it grows, (see experiment 122 on page 125).

Paratonic movements
These are movements which are in response to external stimuli. In most cases only part of the plant is involved but in some cases it involves the movement of the whole plant or organism.

This group may be divided into several sections:—
    a. tropisms
    b. nastisms
    c. tactic movements
    d. hygroscopic movements
    e. turgor movements

TROPISMS

A **tropism** is a growth movement in response to an external stimulus. As the plant organ grows so it curves, the direction of curvature being determined by the direction from which the stimulus originates. Plants respond to various external stimuli and in each case the tropism is given a specific name.

    1. **geotropism** . . . . . stimulus is gravity
    2. **phototropism** . . . . . stimulus is light
    3. **thigmotropism** . . . . . stimulus is contact
    4. **hydrotropism** . . . . . stimulus is water
    5. **chemotropism** . . . . . stimulus is some chemical substance

When referring to the response made by the plant organs, to the various tropisms, certain prefixes are used before the name of the response. If the organ grows in the line of the stimulus, either towards or away from the stimulus then the prefix **ortho-** is used. If the direction of this growth is towards the source of the stimulus it is said to be a **positive response** if it is away from the source of the stimulus then it is a **negative response.** If the plant organ grows across the line of the stimulus at right angles to the stimulus the prefix **dia-** is used, if at an angle other than a right angle then the prefix **plagio-** is used.

When other conditions are constant then the response depends, within limits, on the intensity of stimulus. The lowest value for intensity of stimulus which will produce the minimum observable response is referred to as the **threshold value.** If the intensity of the stimulus is increased, a point is eventually reached when increase in intensity causes no further increase in response. This point is referred to as the **maximum value** of stimulus. In most cases it may be said that as the intensity of the stimulus is increased so the response increases, at first rapidly but later more and more slowly until the maximum value is reached. In some cases the response is constant and the maximum value of response takes place as soon as the threshold value is reached. The **presentation time** is the minimum time for which the plant must be exposed to the stimulus to produce a perceptible response.

Mechanism of the response to light and gravity
The theory which is used to explain the growth curvatures in response to gravity and light is termed the **Auxin theory.** (The activity of auxin in regulating plant growth is dealt with in section 6, pages 99-101, where the action

of other growth regulating substances is also dealt with.).

There are present in most higher plants certain naturally occurring substances which promote growth. Amongst these is heteroauxin, β-indolyl acetic acid (IAA).

In 1931 Kögl and Haagen-Smit isolated, from human urine, a substance which, when placed eccentrically on decapitated coleoptiles, caused marked growth curvatures of the said coleoptiles. The substance was named auxin a, auxentriolic acid. Later another similar substance was isolated from malt, this was called auxin b, auxenolinic acid. A third substance was also isolated, this time from human urine, this substance was called heteroauxin. Since then heteroauxin, IAA, has been positively identified as a naturally occurring substance in plant material, (Kögl and Kosternmans in 1934 working with yeast, and Haagen-Smit et al in 1946, positively identified it in higher plant tissue), but repeated attempts to demonstrate the presence of auxin a and auxin b in plant material have been unsuccessful. It is generally believed that IAA is the naturally occurring auxin, principally responsible for plant growth movements, although there are other naturally occurring compounds within the plant, which have auxin type properties.

Auxin is synthesised at the growing points from an auxin precursor, which is believed to have originated in the storage organs of the seed or in the green leaves. It is passed back from the apex to the region of elongation, which is immediately behind the apex, and in which it is active in controlling the extent of cell elongation. Auxin transport is usually along the longitudinal axis of the plant from apex to base. Movement of auxin in the opposite direction (base to apex) is apparently impossible. The reason for this apparent polarity is not completely understood. It is also possible for auxin to move laterally, across the organ.

The growth curvatures of a shoot or root, produced in response to the stimuli of gravity and light have been found to be due to the asymmetrical distribution of auxin within the organ concerned, but it is still uncertain as to exactly how this asymmetrical distribution occurs.

Much research work is being carried out at present concerning this and other theories to determine exactly, the mechanism of the response of plants to stimuli such as light and gravity.

GEOTROPISM

Geotropism is the directional orientation of part of the plant by a growth curvature in direct response to the stimulus of gravity. Upright stems and tap roots are commonly ortho-geotropic (growing in the line of the stimulus). Lateral stems and roots are either plagio-geotropic or dia-geotropic, while runners and rhizomes are commonly dia-geotropic. Root hairs are insensitive to the stimulus of gravity and are therefore said to be ageotropic.

The auxin theory and the geotropic response of a root

If the radicle of a broad bean seedling is marked at 1mm intervals and the seedling is then placed in a normal growing position, that is with the radicle pointing vertically downwards, then the radicle continues to grow downwards. At the end of the experiment the distances between some of the marks is greater than the original 1 millimetre distances, and by inspection it is found that the growth is confined to a region a few millimetres long just behind the root cap, namely the region of elongation. This is, in fact, the only part of the root which can grow in length. If a similarly marked seedling is placed so that the radicle is horizontal to the line of force of gravity, it is found that as the radicle grows so it bends towards the line of force of gravity until it is pointing vertically downwards, after which it continues to grow in this direction. On inspection of the radicle it will be seen that the bending is confined to the region of elongation and is therefore a growth movement.

In the root an increase in auxin concentration above the optimum inhibits cell elongation, while a decrease below the optimum promotes cell elongation. The auxin passes back from the tip to the region of elongation, through which it can pass in one longitudinal direction, that is from tip to base, (see above). In the vertical root auxin passes back from the tip evenly all round the root and in the region of elongation the auxin is evenly distributed, so the root continues to grow downwards. When the root is placed horizontally it is found that the auxin becomes unevenly distributed in the region of elongation and that there is an increase in the concentration of auxin towards the lower surface, thus towards the upper surface the auxin concentration is below the optimum and there is rapid and extensive elongation of the cells, while towards the lower surface the auxin concentration is above the optimum, elongation is therefore inhibited and is very slow. Since the rate of growth towards the upper surface is greater than that towards the lower surface the root bends downwards.

It appears that, in roots placed at an angle to the line of force of gravity, the normal pattern of auxin translocation is modified so that the auxin accumulates at the lower surface, inhibiting growth in this region which results in the bending movement. Once the root has curved and the tip is again facing downwards the auxin is distributed evenly throughout the root tip and passed back evenly on all sides of the region of elongation and the root continues to grow vertically downwards.

The uneven distribution of auxin occurs initially within the root tip where the auxin tends to accumulate towards the lower surface. It is then passed back unevenly to the region of elongation. It is therefore the root tip which is perceptive to the stimulus of gravity although the effect is seen in the region of elongation.

The auxin theory and the geotropic response of a stem

With respect to growth, stems differ from roots in two important ways. Firstly in stems the region of elongation is longer than in roots since it is continuous through several internodes, and secondly the optimum concentration of auxin for growth is higher in stems than it is in roots. As a result of these differences stems are negatively geotropic while roots are positively geotropic.

In a stem placed vertically in the line of force of gravity the auxin is evenly distributed but when the stem is placed horizontally there is an accumulation of auxin towards the lower surface. In the region of elongation the decrease in concentration which occurs towards the upper surface inhibits growth and in this region cell elongation is slow but towards the lower surface where there is an increase in auxin concentration, elongation is rapid and extensive. As a result of the difference in growth rates on either side of the stem, as it grows so it curves upwards until the apex is vertical, in which position the auxin distribution is unaffected by gravity and auxin is evenly distributed all round the stem.

It appears that it is the tip of the stem which is the seat of perception of the stimulus. As in roots the effect of the stimulus is seen in the region of elongation, although the unilateral distribution of auxin is established in the tip.

PHOTOTROPISM

Phototropism is the directional orientation of part of the plant by a growth curvature in response to the directional stimulus of light. Most stems are positively phototropic. Upright stems are positively ortho-phototropic, as are upright leaves. Lateral stems such as runners and creepers, and leaves which are held at right angles to the light, are dia-phototropic. Roots are, with a few exceptions, insensitive to light. One plant which grows in this country and which has roots which are sensitive to light is ivy, whose adventitious roots grow away from the light into cracks and crevices in walls or tree trunks and are therefore negatively ortho-phototropic. This growth away from light is also demonstrated by the roots of mustard seedlings.

The auxin theory and phototropic response in stems

The response of stems to the directional stimulus of light may be explained by considering the distribution of auxins in the tip and the way in which they are affected by light.

Auxins are formed in the tip and transported to the region of elongation where they are active in controlling cell elongation. Auxins can move in one longitudinal direction only, that is from the apex to the base of the stem, (although some lateral movement across the stem can take place). It is believed that the auxins are in an inactive form when they are passed back from the tip and that they are converted to an active form in the elongating cells. The growth curvature in response to light is caused by the unequal growth of the opposite sides of the stem, this occurs because of unequal distribution of auxin. It has been proved by experiment, that in a stem which has been evenly illuminated on all sides the auxin is evenly distributed around the tip and is similarly evenly distributed in the region of elongation. If a stem is grown in the dark, the amount of auxin produced is increased and it is still evenly distributed all round the tip. In stems a high auxin concentration promotes growth and since the concentration of auxin is increased, the amount and speed of elongation in a plant grown in the dark is also increased. It has been proved that, when a stem is illuminated from one side only, there is an asymmetrical distribution of auxin in the region of elongation. The amount of auxin present in the side away from the light being greater than the amount present in the side nearest the light, elongation is quicker and greater on the shaded side than on the other side and bending occurs, the bend being such as to bring the stem into a position where it is evenly illuminated all round the tip, in which situation the auxin becomes evenly distributed on all sides, the bending movement ceases and the stem continues to grow into the light. The curvature is known to be caused by the unequal growth of the opposite sides of the stem, which occurs because of the unequal distribution of auxin in that region. Various theories have been put forward as to how this asymmetrical distribution of auxin occurs including the following:—

1. that it is due to lateral translocation of auxin in the apex, this lateral translocation being stimulated in some way by light.
2. that it is caused by the inactivation of the auxin by light.
3. that it is the result of the inhibition of auxin synthesis in the presence of bright light.

Some lateral translocation of auxin can occur in the presence of light of low intensities; the reasons why are not clear, but it has been suggested that auxin may move laterally in response to differences in electrical potential between the lighted and unlighted sides of the coleoptile. It has been shown that upon exposure to unilateral light, the shaded side of the coleoptile becomes electrically positive with respect to the side which has been illuminated and the auxin, being acid, is attracted to the more positive shaded side. It is believed that the unequal distribution of auxin and the subsequent curvatures which occur in conditions of low light intensity are the result of the lateral translocation of auxin, although the mechanism by a means of which this takes place is not fully understood.

When considering the response of the plant to light of greater intensity such as normal sunlight, then the asymmetrical distribution of auxin and the growth curvatures which occur because of this are believed to be caused by some other factor in addition to lateral translocation, either the inactivation of the auxin in the presence of light or the inhibition of auxin synthesis in the apex by light. It is possible that all three mechanisms operate.

But irrespective of the means by which this uneven distribution of auxin takes place, it is established that it takes place in the tip of the coleoptile.

THIGMOTROPISM

Thigmotropism is the directional curvature of part of a plant in response to the directional stimulus of contact with a solid object. It is a growth movement. It may also be referred to as **haptotropism**. It is exhibited by twining stems and the tendrils of climbing plants. When the tip of a twining stem comes into contact with a solid object it induces a response which results in the plant organ circling around the support. This is due to a pattern of uneven growth around the circumference of the stem. The rate of growth in any one place is continually changing rhythmically. It is probably a further development of the phenomenon known as **circumnation** which is an autonomic movement characteristic of the shoots of higher plants (see page 125). The growth pattern is quite simple, the sector of the stem opposite the one which is in contact with the support experiences an increase in rate of growth. Thus each sector of the stem in turn experiences a phase of increased growth as it comes into a position opposite the sector which is in contact with the support. This causes a curve to be formed, the concave side of which is next to the support and the centre of which is the centre of the support. The helical action continues as the region of contact with the support and the region of increased growth continually change. As it grows so the stem curls around the support in a tight helix. Not all plants can twine around the same sized support, some require a thin support others a thick one. The thickness of the support depends on the difference between the rates of growth on the two sides of the stem, for example Passiflora gracilis (passion flower) tendrils will twine around a piece of silk while those of Vitis (vine) require a support which has a diameter of about 2 or 3 mm. The stems coil around the support tightly, the radius of the coil being fractionally less than that of the support to ensure a tight fit. The cause of this response and the mechanism involved are as yet unknown. This phenomenon is also exhibited by the tendrils of climbing plants.

HYDROTROPISM

Hydrotropism is the directional orientation of part of the plant by a growth movement in response to the directional stimulus of water. Shoots are insensitive to the stimulus of water but roots are strongly affected by it. Roots are positively hydrotropic, that is they grow towards a source of water. The stimulus of water is greater than that of gravity and roots will break away from gravity to grow towards water. The hypocotyl of some seedlings shows a negatively hydrotropic response. The cause and nature of the mechanism of the response is again unknown.

CHEMOTROPISM

Chemotropism is the directional orientation of part of a plant by a growth movement in response to the directional stimulus of a chemical substance. This is demonstrated only by a few plants.

The growth of the pollen tubes through the tissue of the stigma in a flower is initially a negative response to the stimuli of air and light, as it grows away from these into the tissue of the stigma. As the pollen tube grows down through the tissue of the stigma there is a positive chemotropic response and its growth into the micropyle of the ovule is considered to be under the influence of a chemical substance produced by the embryo sac. The synergid cells are classically described as giving a chemical substance which attracts the pollen tube; some doubt is placed on the synergid function, it may represent part of the archegonium. The growth of fungal hyphae into and throughout the substrate from which they obtain their food substances is an example of a chemotropic response. It is a positive response since they grow towards the substrate. The growth of the haustoria of parasitic plants through the tissue of the host plant, towards the xylem and phloem of the host, is also described as a chemotropic response.

NASTISMS

A **nastism** is a plant movement in response to a stimulus, the response being independent of the direction of the stimulus. The stimuli are commonly non-directional but even in the case of shock movements where the stimulus is directional, the direction of response is still independent of the direction of the stimulus. In some cases these movements may involve growth, in others they are due to sudden changes in the turgidity of the cells. An example of a directional stimulus is that of touch. Non-directional stimuli are sometimes referred to as diffuse, the stimulus being a general change in some condition in the plant's surroundings, such as a rise or fall in temperature or a change in light intensity. Many of the nastisms are inter-related and in some cases it is difficult to determine which is the main stimulus. As in the case of the tropisms each response is given a specific name depending on the stimulus; they are:—

a. **photonasty** . . . . in response to a change in light intensity
b. **thermonasty** . . . . in response to a change in temperature
c. **hydronasty** . . . . in response to a change in humidity
d. **seismonasty** . . . . shock movement in response to the stimulus of touch
e. **chemonasty** . . . . in response to a chemical stimulus
f. **nyctinastisms** . . . . these are day and night movements which are due to a combination of stimuli

NYCTINASTISMS

Nyctinastisms are diurnal night and day movements, such as the opening of flowers during the day and their closing again at night. They are sometimes referred to as sleep movements. Amongst these are also the opening and closing movements of the compound leaves of some plants such as clover and mimosa. The nyctinastisms occur in response to a combination of several stimuli and it is impossible in all cases to say which is the most important. The most obvious of the stimuli concerned with the nyctinastisms is the change in light intensity, but as day passes to night and vice versa there is also a change in temperature and a change in humidity and in most cases it is difficult to determine which is the main stimulus. Nyctinastisms may be growth movements, for example the opening and closing of tulip flowers, or they may be due to a change in the turgidity of a group of cells, for example the collapse of the clover leaf at night.

As mentioned above, the tulip and crocus flowers open at the beginning and close at the end of the day by means of a growth movement. The opening movement is caused by an increase in the rate of growth of the inner surface of each petal causing it to bend outwards. This is later followed by an increase in the rate of growth of the lower surface until the two rates even out. At night the rate of growth on the lower surface is greater than the rate on the upper surface and the flower closes until the two rates even out. It is most likely that this is mainly a response to a change in temperature since if a closed flower is moved from a well lit cold room to a well lit hot room, it opens almost immediately. It is possible that in its natural surroundings the changes of light intensity and humidity may also play a part. In the case of the dandelion flower and other composites the opening and closing movements depend more on changes of light intensity than on changes in temperature. In a number of plants the flowers open in the evening and close during the day; most of these flowers are pollinated by night flying moths. In some of these evening opening flowers it has been shown that the stimulus responsible for the opening is an increase in humidity.

The leaves of a number of plants also show nyctinastic movements, amongst which are clovers, oxalis and mimosa. These plants have compound leaves, and at the base of each leaflet is a pad of parenchymatous water storing tissue referred to as a pulvinus. During the day the cells of the pulvinus are turgid but at night they become flaccid and the leaf collapses. It is a change in the turgidity of the cells of the pulvinus which causes the opening and closing movements of the leaf.

CHEMONASTY

Chemonasty is the curvature of part of a plant produced in response to a chemical stimulus, the direction of curvature being independent of the direction of the stimulus. This response is demonstrated by some insectivorous plants, for example Drosera rotundifolia, Dionaea muscipula, Pinguicula vulgaris.

The leaf of Drosera rotundifolia (round leafed sundew) has a large number of glandular hairs on its upper surface. In the centre of the leaf they are short and erect but towards the margin they are longer and tend to bend outwards. They have a terminal knob-like group of glandular cells which secrete a fluid containing proteolytic enzymes. These enzymes digest the bodies of any insects coming into contact with the tentacles. If the tentacles are stimulated by a liquid containing a nitrogenous substance the tentacles bend inwards and there is an increase in the secretion of digestive fluid. The same response is also produced by some salts, particularly by sodium and ammonium salts and most acids. This curvature is due to a difference in the rate of growth on the two sides of the tentacle. Growth is at first rapid on the side nearest the margin, but after several hours an increase in the growth rate on the other side results in the straightening of the tentacle, which should be completely straight after about twenty-four hours. There is a permanent increase in length of the tentacle. If the tentacle is continuously stimulated the response becomes less and less each time, until eventually there is no response at all.

The leaf of Dionaea muscipula (Venus's fly trap) consists of a flattened leaf stalk with a terminal bilobed lamina, the two lobes of which are normally held at an angle to each other, the angle being a little larger than a right angle. On the margin of each lobe is a row of rigid spike like projections and on the upper surface of each lobe there are three fine spines which project at an angle to the leaf surface. The upper surface of each lobe also has a large number of short glandular papillae. There are also octofid papillae on the under surface of the leaf lobes and on the petiole and simple hairs are present on the under surface of the leaf lamina. When stimulated by a nitrogenous substance the glandular hairs produce a secretion. This stimulus also induces slow movement of the lobes towards each other, the mid-rib acting as a hinge.

At the base of each fine spine like hair is the end of a vascular bundle of the lamina. In the angle, between the base of each spine and the leaf surface, is a group of thin walled cells which are normally distended with water. If an insect touches one of the spines then the spine is pressed against these cells shooting the water into the adjacent vascular bundle, setting up a wave reaction, stimulating the leaf so that the two lobes snap together rapidly. This sudden rapid closing can be regarded as a response to the stimulus of touch. A similar response is shown when the spines are touched by an inanimate object. If an insect is trapped in this way then the lobes will continue to move together slowly, until they interlock tightly along their margins. The glandular hairs then produce their digestive secretion. These are responses to the chemical stimulus of the insect. The lobes remain closed until the insect has been digested and the products absorbed.

In both of these examples part of the plant is moving in response to a chemical stimulus but the direction of movement is not related to the direction of the stimulus.

SEISMONASTY

Seismonasty is the movement of part of a plant in response to sudden stimulation by touch. These movements are referred to as shock movements. There is no connection between the direction of movement and the direction of the stimulus.

One of the best examples of this type of movement is the leaf of Mimosa pudica (a South American leguminose plant). In this case it is not a growth movement but is brought about by a change in turgor in groups of cells, referred to as pulvini, located at the base of each leaflet. Each pulvinus consists of a group of water storing parenchymatous cells. Under normal conditions when the leaf is open, the cells of each pulvinus are fully turgid but if the leaf is suddenly touched there is a rapid change in the turgor of the cells and the leaf collapses. This collapse brings about a remarkable change in the appearance of the leaf. The collapse occurs very quickly but progressively from the point at which the leaf is touched.

Leaves of Mimosa pudica plant before stimulation seen from above to show open leaves

Leaves of Mimosa pudica plant a few minutes after stimulation by touch, side view

TACTIC MOVEMENTS

The plant movements considered on the previous pages, that is the various tropisms and nastisms, have all been movements of part of a plant which is itself in a fixed position. Some of the minute lower plants are however capable of moving about within their environment, an example of these being the unicellular alga Chlamydomonas. The movements of the sperms of the alga Fucus and of the sperms of ferns are other examples of tactic movements.

These movements are usually directional, that is the organism moves either towards or away from the stimulus.

PHOTOTAXIS

This is the movement of a complete organism in response to the stimulus of light.

The green unicellular alga Chlamydomonas is attracted towards light. If a glass jar containing water with plenty of Chlamydomonas dispersed in it, is placed in a position in which it is illuminated from one side only, it will be seen that the green colouration of the water, which is due to Chlamydomonas organisms, will become deeper on the side which is illuminated showing the accumulation of the Chlamydomonas individuals on this side. If however the light is very bright they move away from the light taking up a position, probably in the middle of the container, favourable for photosynthesis. This is a phototactic response.

CHEMOTAXIS

This type of movement is shown chiefly by bacteria and mobile gametes. Bacteria usually show a positive chemotactic movement in response to substances which constitute their food materials. They swim from a region of low concentration to a region of high concentration. They show a negative chemotaxis to many toxic substances.

Another example of this is the movement of fern sperms. Pfeffer showed that fern sperms were strongly attracted to malic acid. Pfeffer found that when a drop of water containing fern sperms was placed on a glass slide under a microscope, and the tip of a capillary tube containing a solution of malic acid was placed in the drop of water, then the sperms could be seen (through the microscope) to be swimming towards the entrance of the capillary tube. It is believed that this is the way in which the fern sperms are attracted to the archegonia containing the eggs, since the archegonia produce malic acid which being soluble in water diffuses into solution around the neck of each archegonium. (Water being essential for fertilisation of the ova in ferns) The spermatozoids swimming in the neighbourhood of the archegonium are attracted towards the neck of the archegonium entering and fertilising the oosphere.

The spermatozoids of mosses are sensitive to sugar.

HYGROSCOPIC MOVEMENTS

An excellent example of this type of movement is that of the annulus of a fern sporangium, others being the movements of the peristome of mosses and elaters of liverworts.

Hygroscopic movements of fern sporangia, for example Dryopteris filix-mas sporangia

The Dryopteris sporangium is a flattened structure the walls being composed of a single layer of thin walled cells with a row of specialised cells called the annulus around the edge (see diagram page 118). The cells of the annulus have thickening on their inner tangential and radial walls, while the outer tangential wall is thin. The young sporangia are protected in a moist atmosphere by the indusium. As the sporangia mature so the indusium shrivels and the water glands on the stalks of the sporangia cease to secrete and the sporangia are exposed to a dry atmosphere. Under these conditions the cells of the annulus lose water due to evaporation. As each of the cells of the annulus loses water so its volume decreases. Since the thick walls of each cell are firm and the thin outer wall is elastic, the outer wall is drawn inwards drawing the thickened radial walls together. As this continues so the thinner parts of the sporangium wall break and the sporangium starts to open slowly. The water within the annulus cells is held together by cohesion. As the drying continues a terrific strain is placed on the cells and the water within them. As the cells contract so a point is reached when the cohesive forces between the water particles break down, the water vaporises and at the same time rapidly increases in volume. The cells of the annulus revert rapidly to their original shape and the annulus and the rest of the sporangium wall snaps back into the original position. Some spores are dispersed as the sporangium opens but it is by the snapping back action that most of the spores are dispersed.

Transverse section of Dryopteris filix-mas pinnule through a sorus, showing mature sporangia

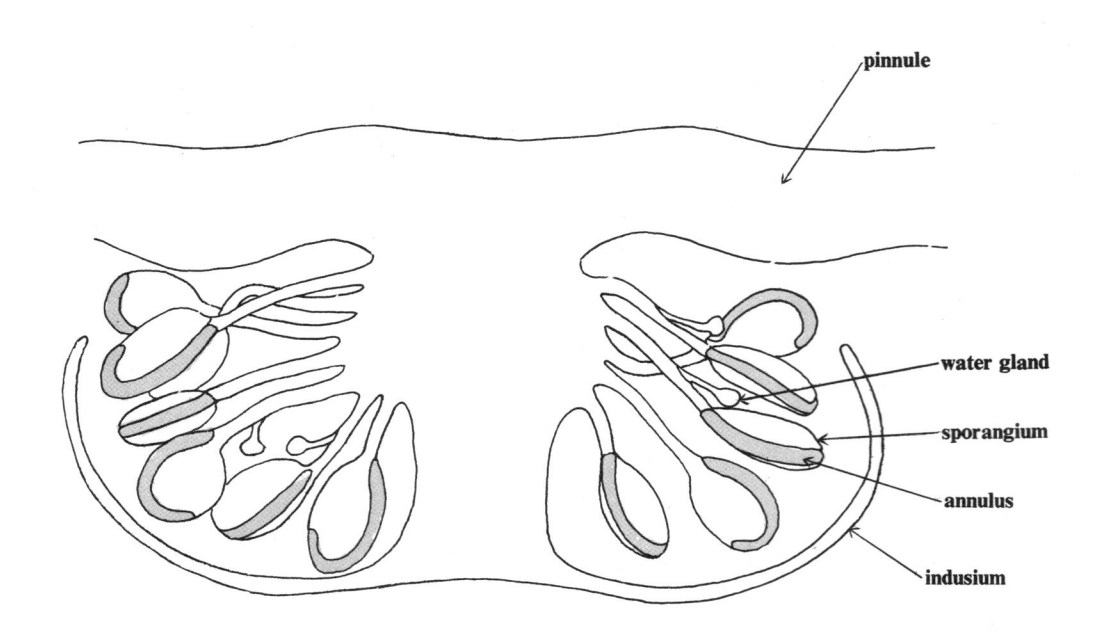

- pinnule
- water gland
- sporangium
- annulus
- indusium

Mature sporangium just before splitting open

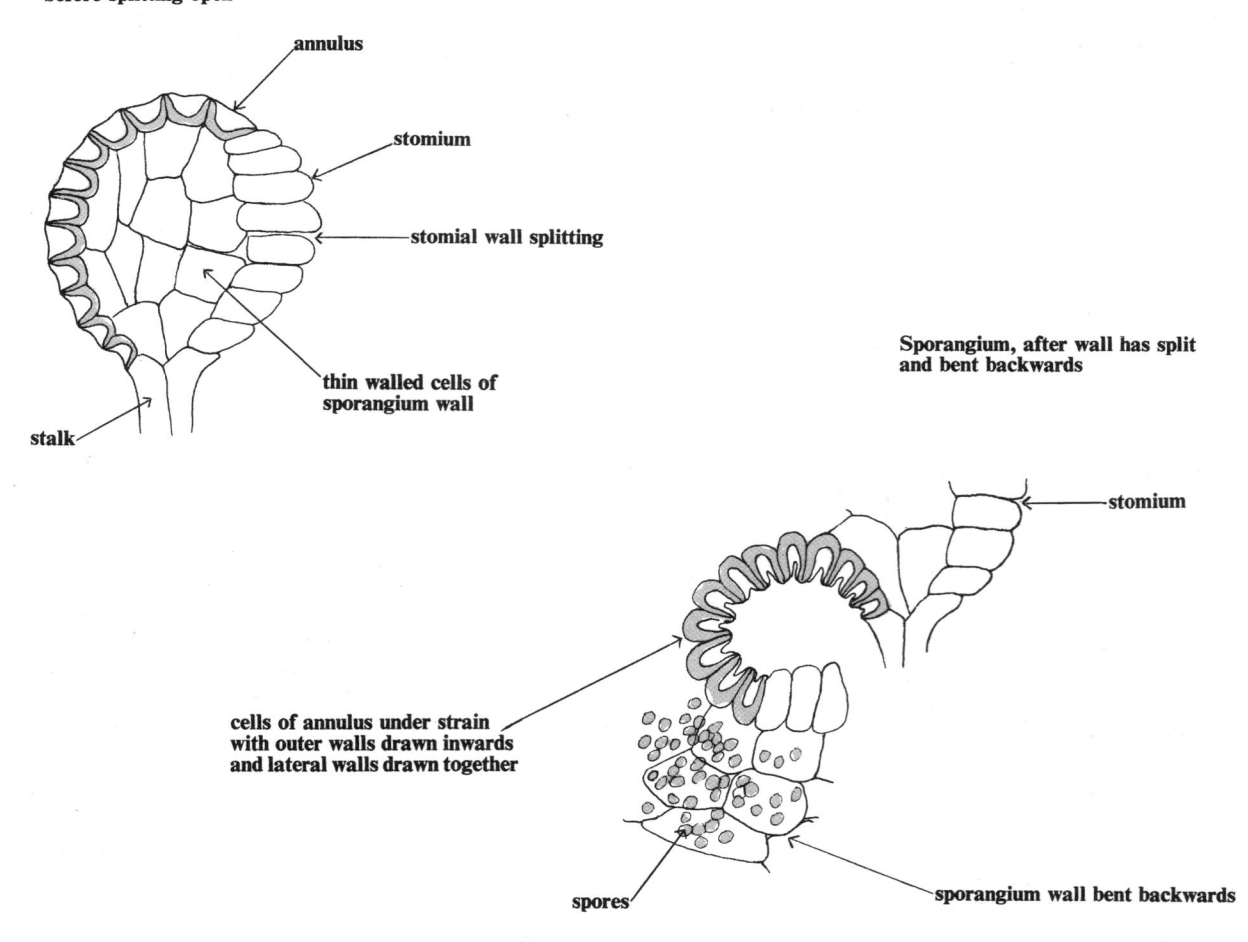

- annulus
- stomium
- stomial wall splitting
- thin walled cells of sporangium wall
- stalk

Sporangium, after wall has split and bent backwards

- stomium
- cells of annulus under strain with outer walls drawn inwards and lateral walls drawn together
- spores
- sporangium wall bent backwards

Hygroscopic action of the peristome of mosses, for example the peristome of Mnium hornum

The peristome is formed in the tissue of the operculum at the apex of the sporogonium. It consists of two sets, each of about sixteen teeth (inner and outer sets) which are composed of strips of specialised cell wall. The operculum is eventually detached.

The peristome teeth are formed in a dome shaped layer of cells, the outer and inner walls of which are cutinised, although the radial walls are non-cutinised cellulose. Eventually the contents and the radial walls disappear leaving the outer and inner cutinised walls which then split radially into outer and inner rings of teeth. These teeth are hygroscopic. In damp weather the outer peristome teeth bend inwards covering the inner peristome and preventing the spores from escaping. In dry weather both rings of teeth bend outwards enabling the spores to disperse.

Mnium sporogonium

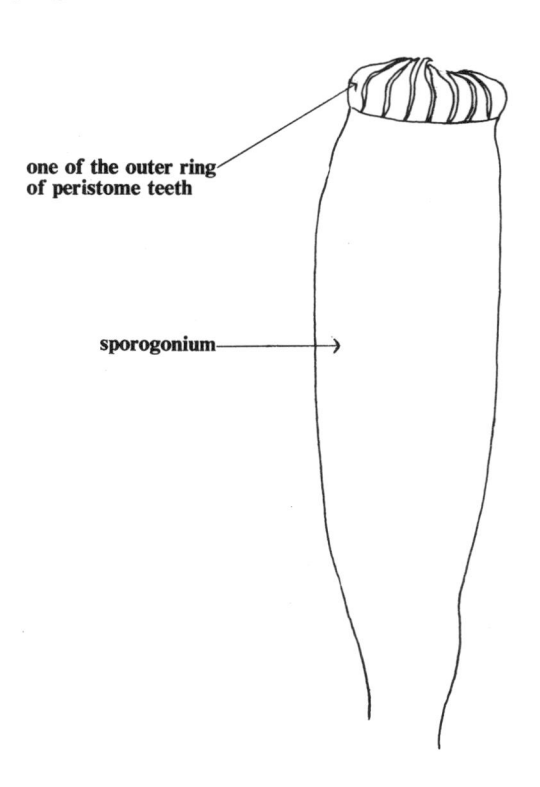

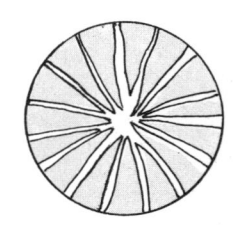

one of the outer ring of peristome teeth

sporogonium

View from above showing outer ring of peristome teeth, the inner ring of peristome teeth are hidden by the outer teeth and are not shown in the diagram

Elaters of liverworts, for example the elaters in Pellia

These are formed within the developing capsules from some of the archesporial cells, which do not develop into spore mother cells. The elaters develop as long thin cells, pointed at both ends with spiral ribs of cutin on their inner walls. When the capsule is mature and bursts then the elaters, which are hygroscopic, react to the atmosphere surrounding the capsule expanding and contracting according to the atmospheric humidity. As they expand and contract they twist and turn rapidly helping to disperse the spores

TURGOR MOVEMENTS

These are responsible for the opening and closing movements of stomata (see page 11 and Book 1 page 31).

Loss of water from thin walled cells within a plant can cause wilting either of part of the plant, such as the wilting of leaves and leaf and flower stalks, or the wilting of the complete plant.

AUTONOMIC MOVEMENTS

These are spontaneous movements. They include the movements of chromosomes during nuclear division, see page 98; and cyclosis which is the name given to the streaming movement of the cell contents, see experiment on page 125. Nutation or circumnutation is another autonomic movement and is demonstrated in the experiment on page 125.

EXPERIMENTS TO DEMONSTRATE TROPIC RESPONSES

Method of growing broad bean seedlings for use in experiments
Before planting, the broad bean seeds should be soaked in water for about twenty-four hours. They must be planted carefully to ensure that the radicles and plumules grow in a vertical direction. They may be grown in soil, or potting compost, which must be of small evenly sized particles to prevent deformity of the plumules or radicles. They may also be grown by pinning the soaked seeds to a piece of cork or soft wood which is then placed upright in a jar in the bottom of which there is a little water. The seeds should be surrounded by a little moist cotton wool. The jar should be covered with a glass disc to maintain a moist atmosphere within.

Method used for marking the radicles of broad bean seedlings for use in experiments
The broad bean seedlings should be used about 4 to 6 days after planting when the radicles should be about 1 inch long. The **radicle** should be marked from the tip backwards with transverse lines made in indian ink at 1mm intervals. One of the easiest ways is with a piece of cotton which has been placed on an inked pad. A mapping pen may be used if it has a very fine nib. **Shoots** of seedlings should be marked at 10 mm intervals, from the tip backwards over several internodes.

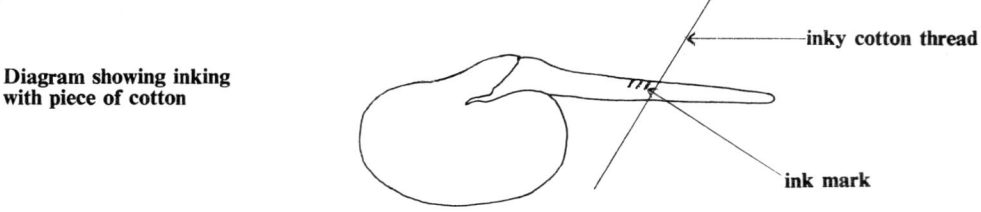

Diagram showing inking with piece of cotton

Experiment 112

To demonstrate the effect of gravity on a growing root

APPARATUS REQUIRED
gas jar with glass cover or cork
blotting paper and pins
piece of soft wood, such as balsa wood, or a piece of cardboard
2 broad bean seedlings, at a stage in development at which the radicles are about 1 inch long (4-6 days after planting)

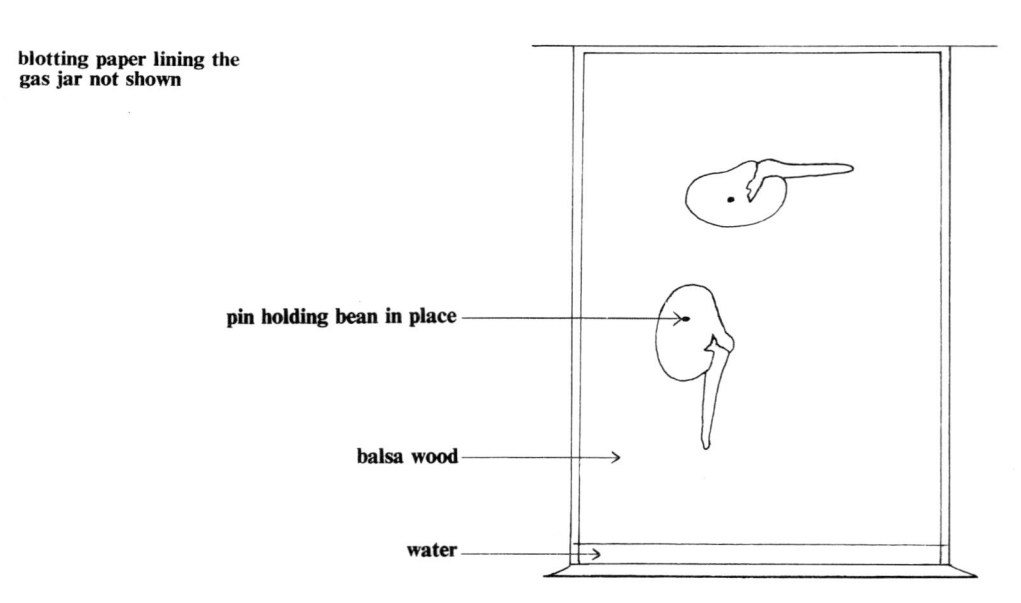

blotting paper lining the gas jar not shown

METHOD OF PROCEDURE
Line the gas jar with a piece of blotting paper. Place a little water in the bottom of the jar and cover the top with the glass cover. Leave for a short time to ensure that the atmosphere inside the jar is moist. Cut the piece of wood (or cardboard) so that it just fits when placed upright inside the jar. Take the young bean seedlings and mark the radicles carefully at 1mm intervals as described above. Attach the seedlings to the wood with pins through their cotyledons. Arrange the seedlings so that the radicle of one is pointing vertically downwards, while the radicle of the other is horizontal (as in the diagram). Place the piece of wood with the seedlings attached, into the gas jar making sure that the radicles do not touch the water. Cover the jar with the glass cover and leave for 1 day, after which inspect the radicles noting any change in the distances between the ink marks and any change in direction of growth.

Experiment 113

To demonstrate the influence of gravity on a growing shoot

APPARATUS REQUIRED

young sunflower seedling rooted in a small pot; alternatively you may use a young shoot cut from a plant and inserted through a hole in a cork which is then placed in a test tube filled with water
clamp and stand
piece of glass

METHOD OF PROCEDURE

Take the young sunflower seedling and mark the shoot at 10 mm intervals over several internodes as described on page 120. Then lay the pot on its side so that the shoot is horizontal. Place an upright piece of glass behind the shoot. Mark on the glass, at intervals of about a quarter of an hour, the position of the tip of the shoot. Note any change in direction of growth.

Note: When a shoot bends upwards under the influence of gravity it does not attain a true vertical position at once, but overshoots the vertical several times in either direction before it becomes vertical.

Experiment 114

To demonstrate that the geotropic response of a radicle is a definite growth movement and is not due to wilting

APPARATUS REQUIRED

glass dish with vertical sides
gelatine
young pea seedling
cotton wool and pins
small piece of cardboard

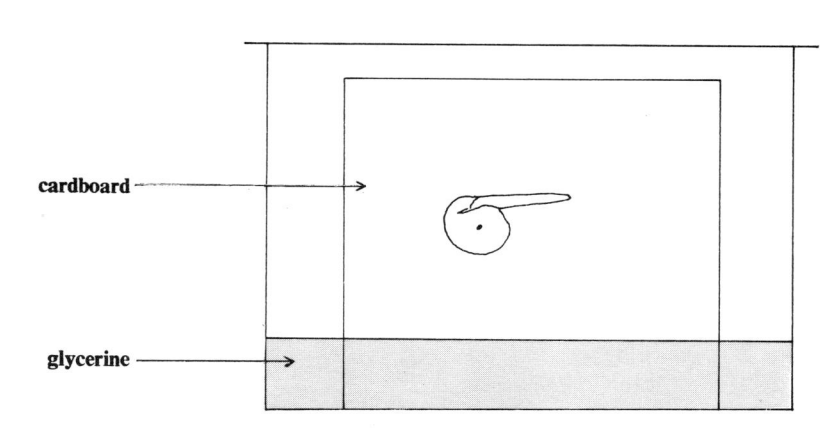

METHOD OF PROCEDURE

Make a 15% gelatine solution and pour it into the glass dish until there is a layer about 2 cm deep. Place the dish on one side to allow the gelatine to set. When the gelatine is nearly solid stand a piece of cardboard upright in the dish. When the gelatine is solid take the seedling, which must have a straight root about three quarters of an inch long, and place it on the gelatine so that the root lies straight and horizontal, as shown in the diagram. Fix it firmly in position by a pin passed through the cotyledons and into the cardboard. Place a piece of damp cotton wool around the seedling. Cover the dish with a glass plate and leave for a few days, then note the direction of growth of the radicle.

Note: If the radicle bends into the gelatine against the opposition of the substance of the gelatine it is doing work and a definite growth movement occurs.

Experiment 115

To demonstrate the effect on the growth of the radicles of broad bean seedlings of neutralising the unilateral influence of gravity

APPARATUS REQUIRED

Klinostat—this is a piece of apparatus designed to neutralise the effect of gravity by keeping the plant in continual rotation about a horizontal axis so that the plant is not in any one position for sufficient time for the presentation time to be reached.
6 broad bean seedlings; they should have straight radicles about 1 inch long
cork disc with transparent cover
cotton wool and pins
clamp and stand

Side view of Klinostat

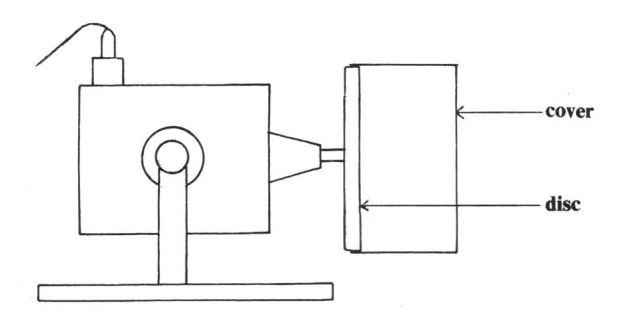

View of beans on disc, cotton wool and cover not shown

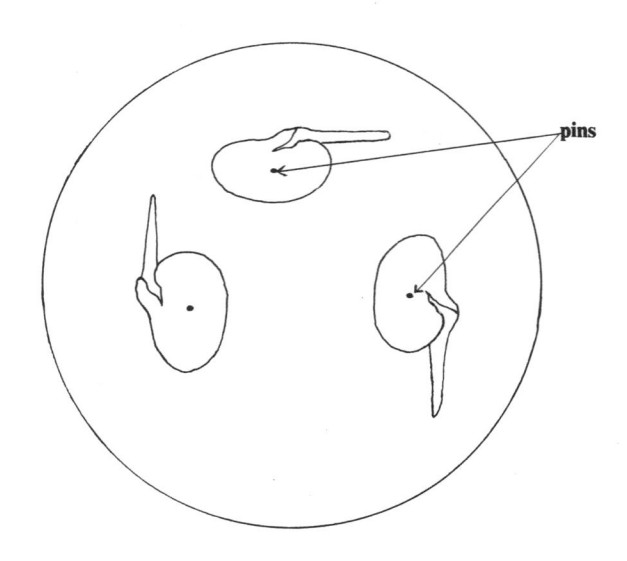

METHOD OF PROCEDURE

Take 3 of the seedlings and cover the cotyledons of each with a thin piece of damp cotton wool then attach each seedling, by a pin through the cotyledons, to the cork disc on the klinostat. The seedlings should be placed so that their radicles are pointing in various directions. A plan should be made to show their positions. Cover the disc with the transparent cover and switch the klinostat on to start the disc rotating. Take the other 3 seedlings and cover each with damp cotton wool and attach it to a cork disc similar to that on the Klinostat. The seedlings should be placed in positions identical with the positions of the seedlings on the klinostat. The disc should be supported by the clamp and stand and covered by a transparent cover. Leave both the disc and klinostat for several hours, after which compare the direction of growth of the radicles of the seedlings on the klinostat with those on the stationary cork disc.

Note: If the klinostat neutralises the unilateral effect which gravity has on the seedlings, then the seedlings which were placed on it will show no curvature. Your results should include diagrams showing the direction of growth of the seedlings at the beginning and at the end of the experiment.

Experiment 116

To determine the presentation time of a radicle of a young bean seedling

APPARATUS REQUIRED

4 young broad bean seedlings with straight radicles about 1 inch long (4-6 days after planting)
piece of soft wood such as balsa wood
clamp and stand
blotting paper, cotton wool and pins
Klinostat

METHOD OF PROCEDURE

Cover the piece of wood, with damp blotting paper, and support it in a vertical position by the clamp and stand. Take 1 of the seedlings, cover it with a thin piece of damp cotton wool and attach it by a pin passed through its cotyledons to the piece of wood. The seedling should be placed so that the radicle is horizontal to the line of force of gravity. Repeat this at 15 minute intervals until all 4 seedlings are attached to the piece of wood. Leave for a further 15 minutes (1 hour in all since first seedling attached). Remove all 4 seedlings from the wood and attach them to the klinostat which should then be covered with the transparent disc. Start the klinostat. After 1 hour note which, if any, of the seedlings show curvature of the radicle, and from your results determine the presentation time for a young broad bean seedling.

Note: Those seedlings which were exposed to the stimurus of gravity for a time interval which was longer than the presentation time will show curvature of the radicle in response to the stimulus. Those exposed for less than the presentation time will obviously show no response.

When the seedlings are placed on the piece of wood they should be marked in some way for later identification.

Experiment 117

To demonstrate the effect, on the growth of a young shoot, of neutralising the unilateral influence of gravity

APPARATUS REQUIRED

Klinostat fitted with flowerpot holder
2 young sunflower or broad bean seedlings in pots (one of the pots must fit into the holder on the klinostat)
clamp and stand

Side view of Klinostat fitted with flower pot holder and seedling

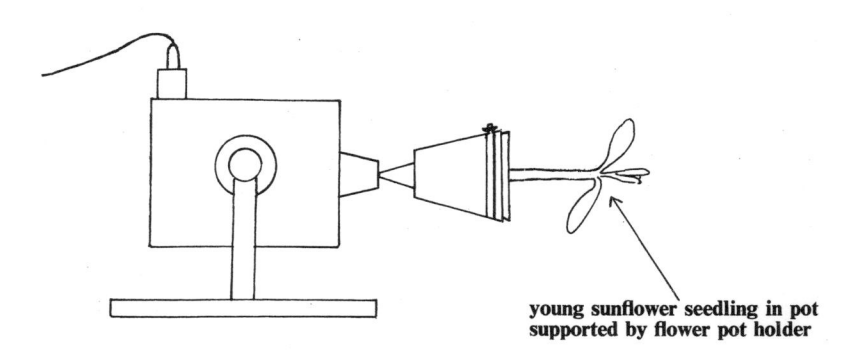

young sunflower seedling in pot supported by flower pot holder

METHOD OF PROCEDURE

Fix one of the pots containing the seedlings into the holder on the klinostat so that the pot and seedling are held in a horizontal direction. Then switch on the klinostat. Support the other pot with the clamp and stand so that the pot and seedling are also held in a horizontal position. Leave both seedlings in this position for several hours. Then note the direction of growth of both seedlings and any curvature which may have taken place.

Care should be taken to ensure that the seedlings are illuminated evenly from all directions, thus avoiding any phototropic curvature.

Experiment 118

To demonstrate hydrotropism in young bean seedlings

APPARATUS REQUIRED
porous pot
large flower pot (or bowl) about 6 inches deep
good soil or potting compost
bean seeds which have been soaked in water for at least 24 hours

METHOD OF PROCEDURE
Take the flower pot and partially fill it with damp soil. Place the porous pot in the centre then completely fill the flower pot with soil and, at the same time, plant a ring of broad bean seeds around the porous pot.

Fill the porous pot with water. Leave the experiment for a week to 10 days topping up the water in the porous pot at intervals.

After about 10 days remove the soil from the flower pot, being careful to keep it in one mass, then gently remove the soil from around the edges until the roots of the seedlings are exposed.

Note: If the roots bend towards the porous pot forming a tightly fitting mass around it then they must have been attracted towards the water in the pot.

Experiment 119

To determine which has more influence on the growth of the roots of young seedlings, either water or gravity

APPARATUS REQUIRED
wooden block
fine flat bottomed sieve
sawdust
mustard seeds

METHOD OF PROCEDURE
Take the sieve and place a layer of damp sawdust about 1 inch thick in the bottom of it. Sprinkle a few mustard seeds over the damp sawdust and cover with another thin layer of sawdust. Support one side of the sieve on a wooden block so that the sieve is inclined at an angle to the line of force of gravity. Leave for a few days checking at intervals that the sawdust is damp and noting the direction of growth of the radicles. Note especially any changes in the direction of growth, and the position of the radicles in relation to the damp sawdust when these changes occur. Then determine by the direction of growth of the radicles which has the stronger influence on growth, gravity or water.

Experiment 120

To demonstrate thigmotropism

APPARATUS REQUIRED
young portions of sweet pea or vetch plants with some tendrils which are uncurled or only slightly curved
glass rod
gelatine solution
powdered emery or fine sifted sand

METHOD OF PROCEDURE
Take the glass rod and dip it into gelatine solution, remove and leave it to dry. Select one of the uncurled tendrils and gently stroke the inside of it with the dry gelatine covered rod. Then dip the rod into the powdered emery or sand and gently stroke the inside of a second tendril. A third tendril should be left untouched to act as a control. Leave for quarter to half an hour, then note any curling which may have occurred in the tendrils touched with the glass rod.

EXPERIMENTS TO DEMONSTRATE AUTONOMIC MOVEMENTS

Experiment 121

Simple experiment to demonstrate cyclosis

APPARATUS REQUIRED
shoot of Elodea canadensis in water
microscope
microscope slide

METHOD OF PROCEDURE
Remove a leaf from the Elodea shoot and mount it in a drop of water on a microscope slide. Leave the leaf for a few minutes to become adjusted to its new surroundings and for cyclosis to recommence. Observe the contents of the elongate cells in the midrib position and note the streaming motion of the contents.

Mount a second leaf from the Elodea shoot in liquid paraffin and thus prevent air from reaching the leaf. Note that the movement of the cytoplasm soon stops showing that energy obtained from respiration is required to maintain movement.

Mount a third leaf from the Elodea shoot in a drop of ether water (a few drops of ether in 100 ml of water). Note that the trace of ether stops cyclosis but that it may be re-started by washing the leaf in tap water.

Experiment 122

To demonstrate circumnutation

Note: Circumnutation is an autonomic growth movement. It is the helical movement of the apex of a shoot as it grows. It is caused by differences in the rate of growth around the circumference of a young stem. In the young internodes the rate of growth is not equal on all radii, and the rate of growth in any one sector is continually changing rhythmically so that each sector in turn experiences the maximum growth rate, with the rates of growth building up to and then decreasing after the phase of maximum rate of growth. The sector immediately opposite will be experiencing the minimum rate of growth, which will similarly be experienced by each sector in turn. This phenomenon is highly developed in climbing plants which twine around their support (see also page 114).

APPARATUS REQUIRED
young kidney bean or scarlet runner plant, twining plants are used as they demonstrate this type of movement better than other plants, although it is possible to use a non-twining plant
2 clamps and stands
glass plate
light pointer made from a thread of drawn out glass, fitted with a minute triangle of paper, and with a tiny bead of sealing wax on the tip of the glass thread
small piece of soft wax

Diagram showing pointer

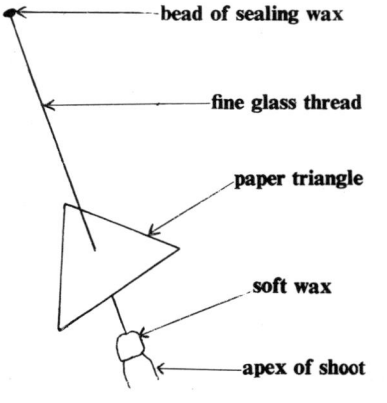

METHOD OF PROCEDURE
Fix the pointer to the apex of the shoot of the potted plant using a spot of soft wax. Arrange the glass sheet above the plant holding it in place by the clamps and stands. Observe the position of the pointer, viewing it so that the bead of wax on the tip of the pointer is in line with the centre of the paper triangle then mark with an inkspot the point on the glass sheet at which the tip of the pointer is pointing at that moment. Number the mark and note the time. Repeat the observation at regular intervals numbering the marks in sequence. At the end of the experiment join the marks in sequence to show the helical path of the apex as it was growing.

Note: It is important to avoid any error which may occur as a result of a phototropic curvature. The plant should therefore be in a position in which it is illuminated evenly from all sides. If this is impossible then it should be surrounded by a roll of black paper and illuminated from above.

SOIL

Soil forms the superficial layer covering most of the earth's crust.

The type of vegetation growing in the soil is directly related to its structure and composition, thus each type of soil supports a characteristic form of vegetation which may vary slightly according to local conditions.

A deficiency in the soil of any of the minerals which are important to plant growth results in the plants being stunted or malformed in some way.

The study of soils (pedology), is obviously linked with the study of ecology.

FORMATION OF SOIL

Soil is formed from rock by a series of processes the most important of which is referred to as **weathering.** As a result of the weathering processes the parent rock is broken into fragments and altered chemically to form the basic raw material, (referred to as the **mineral skeleton**), from which the soil proper is formed. The action of living things on the mineral skeleton, combined with the general effect of the climate results in the formation of mature soil.

Soil may be formed in situ from the rock over which it lies; in which case it may be referred to as a **sedentary soil.** Alternatively the soil may bear no relationship to the underlying rock; the soil being formed from rock which was primarily weathered to form variously sized boulders or small sized particles, which were then transported by some agency such as rivers, streams, flood water or landslides, and secondarily weathered, to form soil, where deposited. These soils may be referred to as **transported soils.**

The type of soil formed depends partly on the nature of the parent rock and partly on the circumstances through which this has passed during the formation of the soil, an important factor being the climatic conditions.

The weathering of rock
The weathering of rock is the result of the combined effects of water, heat and frost on the rock.

THE EFFECT OF WATER ON ROCK
Rain water falling on rock tends to remove small particles from the surface especially if the surface is sloping and the rain is heavy. The particles are either washed over the surface of the rock down to lower regions where they are deposited or washed through cracks and crevices in the rock and carried away in streams and rivers.

As water passes through the joints in the rock it tends to erode the surface of the rock at the joint.

Water can dissolve soluble substances in the rock with the resulting dissolution of the rock. This is a form of **chemical weathering.** It is sometimes called **rock rotting** as distinct from **rock breaking** which occurs as a result of the effect of frost (see below).

Examples of rock rotting:—

1. Rock salt in contact with water either at the surface or below ground is readily dissolved and removed in the water.

2. Limestone, which consists mainly of calcium carbonate, is insoluble in pure water but can be dissolved by rain water which contains a minute proportion of atmospheric carbon dioxide in solution. As rain water, with dissolved carbon dioxide, percolates through the joints in limestone rock, the limestone is dissolved at the surface of the joints, which therefore increase in size, the rock eventually breaking to form variously sized boulders which are liable to rotting in the same way. This rotting can occur at any limestone surface in contact with rain water containing dissolved carbon dioxide. Elaborate land forms may be produced at intermediate stages but there is generally complete dissolution of the rock in the end.

3. Iron compounds occurring in rocks are liable to rusting in the same way as manufactured iron, and rain water passing through joints in rock rich in iron compounds results in rusting of the iron and the consequent softening of the rock faces adjacent to the joints, followed by the breaking of the rock into boulders.

THE EFFECT OF FROST ON ROCK
Rain water drains into cracks in the rock, then, when the temperature falls, the water freezes and in doing so expands, exerting a pressure on the surrounding rock which tends to force the rock apart. This is an example of **rock breaking.** In cold climates many mountainsides are covered with sharp edged angular pieces of rock which are formed in this way, typical of steep slopes where bare rock is exposed to frost, referred to as scree.

THE EFFECT OF HEAT ON ROCK

Heat speeds up the chemical changes occurring within the rock itself. In regions of intense heat the rocks heat up very quickly during the day, often getting very hot, but they cool equally quickly by night. This wide contrast in temperatures which occurs fairly rapidly puts stress on to the rock resulting in the scaling off of thin flakes or the splitting of the rocks into slabs which in turn, because of the effect of the alternating high and low temperatures are also broken down.

Another effect of the action of heat on rock is the effect on the crystals within the rock which expand and contract at different rates depending on their size and colour, thus putting great strain on the rocks eventually resulting in their complete breakdown. This is a form of **chemical weathering,** see below.

CHEMICAL WEATHERING

This involves the decomposition of the original minerals mainly by hydrolytic reactions occurring under acid or alkaline conditions.

Rock rotting (see page 126) is a form of chemical weathering as are the results of the effect of heat on the rock, see above.

So far we have considered the factors which result in the conversion of the parent rock into the primary raw material (mineral skeleton) from which mature soil is formed. There are three factors which are connected with the formation of mature soil:—

1. The nature of mineral skeleton
2. The gradual colonisation of the mineral skeleton by plant and animal life, sometimes referred to as biological weathering.
3. The effect of the climate on the soil as it matures.

Biological weathering

The effect of living things on the developing soil may be referred to as **biological weathering.** The first living things to colonise developing soil are the lowest forms of plant life, that is algae, lichens, liverworts, mosses, all of which help to prepare the soil for higher plants by breaking up soil particles and providing organic nutrients in the form of humus. As dead plant material begins to accumulate so it becomes colonised by bacteria and other micro-organisms the action of which results in the gradual decay of the dead plant material with the resulting formation of a black colloidal substance called **humus,** see also page 128.

As the soil matures so it is gradually colonised by higher plants the roots of which grow down through the soil helping to break up the lower layers. An important factor in the opening up of the lower layers is the effect of tree roots which grow down through minute cracks in the lower layers. As they grow so they increase in size forcing the rock apart.

The effect of the climate on a developing soil

The climate plays an important part in the formation of a mature soil. Certain types of soil have been shown to be characteristic of certain climates. Under the same climatic conditions different types of parent rocks tend to produce the same type of soil while identical parent rocks in different climatic regions will produce different soils.

If the mineral skeleton does not contain a sufficient variety of constituents then a well defined climatic soil type cannot be produced; for example a very pure limestone has a poor variety of constituents and cannot form a typical soil. In this way local variations can occur within any one climatic region. The two climatic factors having most effect on the formation of a mature soil are rainfall and temperature.

RAINFALL

Rain water percolates through the surface layers of the soil carrying with it soluble substances and fine insoluble particles which are deposited in lower levels. The removal of soluble salts is referred to as **leaching,** while the layers from which the fine insoluble particles have been removed are said to be **eluviated.** If rainfall is high there is a tendency for a maximum amount of leaching, if it is low then a minimum amount of leaching. This downward percolation of water through the soil results in the development of layering within the soil, (see page 129, under soil profiles.)

TEMPERATURE

The temperature of the surroundings plays an important part in the development of the mature soil for it affects the rate of evaporation of water from the surface of the soil. As the temperature of the surroundings increases, so the rate of evaporation of water from the surface of the soil increases, thus in very hot climates the rain water tends to evaporate from the surface very quickly and there is less leaching. In a hot climate with a low rainfall there will be very little leaching, while in a cold climate with a high rainfall the soil will be heavily leached. Obviously the final effect depends on the balance between the two. Occasionally the amount of water lost by evaporation is higher than the rainfall, in which case water tends to be drawn up from the lower levels bringing with it soluble salts.

THE STRUCTURE OF THE SOIL

It is difficult to draw a hard and fast line between **topsoil** and the underlying rock, especially in the case of sedentary soils, where the topsoil merges into the rock base through a layer which is referred to as the **subsoil**. This subsoil consists of stones intermediate in size between the soil particles and the boulders of rock.

In a mature sedentary soil one may distinguish three regions, which are:—
    a. topsoil
    b. subsoil
    c. rock base

These three layers are often clearly seen in places where a deep cut has been made into uncultivated soil, for example at the edge of a gravel or sand pit, at the top of a cliff or where excavations are being carried out prior to the erection of new buildings etc.

This layering is not seen in transported soils especially if they have been formed from very fine alluvial deposits, since the rock base over which they lie is not the parent rock.

The thickness of the topsoil varies, and in some cases is governed by the slope of the land, the direction and strength of the prevailing winds and the amount of rainfall; for example a heavy rainfall on to a steep slope will wash the fine soil particles down to lower slopes or into dips and the soil on the steep slope never develops beyond the initial stages and is always very thin.

Soil acidity

When measuring the acidity of the soil one measures the active acidity in the form of the hydrogen ion concentration of the soil. All liquids which have water as one of the constituents contain some free positively charged hydrogen ions and some hydroxyl ions. If there are equal numbers of hydrogen and hydroxyl ions then the liquid is said to be neutral, if there are more hydrogen ions than hydroxyl ions the liquid is said to be acid. An alkaline solution is one in which there are more hydroxyl ions than hydrogen ions. The product of the concentration of the hydrogen and hydroxyl ions expressed in grams is constant at a given temperature, so if one is known then the other can very easily be calculated. It is usual to refer to the hydrogen concentration as grams of active or ionised hydrogen per litre of liquid, but since this can be rather clumsy the term **pH** has been devised, the pH being the logarithm to the base ten of the reciprocal of the hydrogen ion concentration. The hydrogen ion concentration of decinormal hydrochloric acid is 0.1 or 10^{-1} and the pH is therefore 1, that is the index of the hydrogen ion concentration with the negative sign changed to positive. The hydrogen ion concentration of pure water is 10^{-7} therefore the pH is 7.

The pH value for a soil can be fairly easily measured since certain dyes show a definite change of colour according to the pH of the solution, for example litmus solution is red in an acid solution but blue in an alkaline solution. These dyes are referred to as indicators. Their usefulness depends on the range of pH values over which they operate and the distinctiveness of the colour changes. In most cases the colours are only constant for a given concentration of indicator and some can only be used in either acid or alkaline solutions. A whole range of indicators is required to cover the complete range of pH values.

Accurate measurements of soil acidity can be made using indicator solutions with known colour ranges for which purpose there are various soil testing kits which often include solutions for testing for the presence or absence of various minerals necessary for plant growth, (see page 134).

The hydrogen ions "present in" the soil come from soil water containing carbon dioxide in solution, that is weak carbonic acid, and from the organic acids of the humus.

Humus content

Humus is a black colloidal substance which has a high water holding capacity and is capable of adsorbing calcium and other bases preventing their removal from the soil by leaching. It is formed mainly from the lignin and cellulose in dead plant material by the action of various micro-organisms present in the soil, the most important of which are the bacteria.

The majority of soils are freely colonised by bacteria and other micro-organisms involved in the disintegration and decay of dead plant material. Some soils however are unfavourable to most plant and animal life, and are colonised only by specialised plants; here the bacteria and other micro-organisms involved in the formation of humus are absent, and the dead plant material remains on the surface of the soil forming an acid peat. Thus the humus content of the soil can vary from less than 1% up to 100% in pure peat.

Humus has primarily a high degree of acidity which is due to the organic acids which are formed during the decomposition of plant material, unless it is derived from plant material having a high base content. The acidity of the humus can therefore vary, depending on the plant material from which it is derived.

Humus is continually disintegrating and replenishing the mineral nutrients within the soil forming an important source of these substances for higher vegetation.

There are various types of humus. **Mild humus** or **mull** which is produced in soils with a moderate to high temperature, adequate moisture, good aeration and plenty of basic ions, and where there is a large quantity of plant debris which is broken down quickly. In soils unfavourable to plant life and inhabited only by specialised plants decay is slow, basic ions are short and a very acid **raw humus** or **mor** is formed.

Air and water content of the soil
The water content of the soil is a variable factor depending not only on the climatic conditions but also on the physical properties of the soil which determine its capacity for holding water. For example clay, which is composed mainly of fine particles, has high water retaining properties therefore drainage of rain-water is difficult and aeration is poor and as a result, the soil tends to be wet, cold and poorly aerated. At the other extreme a sandy soil is composed of large particles permitting speedy percolation through the soil of any rain-water, it therefore has a low water holding capacity and tends to be dry and well aerated. Obviously the physical properties of the soil and the climatic conditions also govern the ability of the soil to draw water from lower layers, referred to as **soil capillarity.**

Temperature
Dark soil tends to absorb more heat than the light coloured soils therefore the temperature of a dark soil is greater than that of a light soil under similar conditions. Soils with a high water holding capacity and poor aeration tend to be cold and often waterlogged, while those with a low water holding capacity and good aeration are warmer and dryer. The slope of the surface of the soil and whether it is towards the sun or away from it, the degrees of slope and the amount of evaporation from the surface also affect the soil temperature.

TYPES OF SOILS
There are various ways of classifying soils:—
1. according to the type of soil profile
2. according to the soil texture
3. according to some special chemical characteristic

The classification of soils according to the type of soil profile

SOIL PROFILE

As a soil develops naturally from mineral particles and humus it tends to become stratified. This is due mainly to the downward percolation through the soil of rain water, which carries with it fine particles and dissolved salts which are deposited in lower levels. This results in the formation of a number of layers termed **horizons.** These soils may be referred to as **climatic soils** since the type of soil formed is, to a certain extent, dependent on the climatic conditions, that is the amount of rainfall and the temperature of the surroundings. This succession of layers from surface to rock base is referred to as a **soil profile.** Although a particular type of soil may be characteristic of a particular climate the type of soil is not necessarily uniform and constant in any one climatic region, differences being due to variations in parent rock and local conditions.

Cultivation destroys the natural layering of the soil and cultivated land is uniform down to the subsoil.

The three commonest climatic soils in the British Isles are:—
1. brown earths
2. podsols
3. blanket bog peats

Depending on special local conditions such as water/subsoil relationships certain other types of soils are found; these are:—
4. rendzina
5. meadow soils
6. fen peat
7. raised bog or moss peat

1. BROWN EARTHS

These are typically formed in regions where the temperature and rainfall are moderate but the ratio of precipitation to evaporation is such that the calcium carbonate is leached from the upper layers and other soluble salts, such as potassium and magnesium, are also partially leached from the upper layers. These surface layers are however rich in bases including calcium (but not calcium carbonate) and humus, and the soil is very fertile. Characteristic vegetation is deciduous woodland.

Brown earths are typically formed from clays and loams but they are also formed from any rock containing a sufficient variety of mineral elements, such as impure sandstones, limestones and certain metamorphic rocks.

2. PODSOLS

Podsols are characteristic of cooler and wetter regions in the north and west, where the soil is too damp and cold for the thriving of humus forming bacteria and other micro-organisms, thus there is little or no decay of dead plant material which collects as a layer of very raw humus on the surface of the soil. This raw humus is referred to as **mor**. Rainfall is fairly high and as the acid water percolates through the soil, leaching of the upper layers is extreme, these layers becoming deficient in basic salts and often grey in colour. Characteristic vegetation is coniferous forests, moors and heaths.

3. BLANKET BOG PEAT

These soils are formed in regions similar to those in which podsols are formed but where the rainfall is higher. They are formed in places where the surface is usually waterlogged because of poor drainage, such as in a dip, or where rainfall is high and because of the perpetually moist air evaporation is low. There is very limited colonisation of this type of soil since the plants must be able to live in an acid soil which is waterlogged and therefore has an oxygen deficiency. These plants include Sphagnum (bog moss), some sedges, and a few other plants where drier.

4. RENDZINAS

This type of soil is primarily determined by the nature of the parent rock, which is limestone, and not by the climate. It is formed in regions favourable to the formation of brown earth soils and the characteristic vegetation is grassland. Where it is wooded then in the south it is beechwood and in the north and west ashwood. It is a very alkaline soil, the whole profile being saturated with free calcium carbonate. The soil is usually very thin and formed on fairly steep slopes, the soil is deeper on the gentler slopes and on flat surfaces it tends towards the brown earth type.

5. MEADOW SOILS

Meadow soils are formed on flat ground usually from the alluvium of river valleys. They are formed in places where there is a fluctuating water table fairly near the surface and drainage is impeded, hence leaching is also impeded and humus tends to remain near the surface. Meadow soils are rich in bases and very fertile. They are often flooded in the winter this flooding bringing fresh silt which helps to maintain fertility. The characteristic vegetation is grassland. When meadow soils are drained they tend to develop into brown earth soils.

6. FEN PEAT

Peat is a pure organic soil formed where the ground is constantly waterlogged and permanently deficient in oxygen and as a result humus accumulates and does not disintegrate. In regions where the water has drained through calcareous or other basic rocks then **fen peat** is formed as seen in East Anglia. If the surface rises above the level of the ground water, due to the accumulation of plant remains, then aeration is increased and the soil can support a greater variety of plants. Where the climate is humid and the peat, being above the level of the groundwater, is no longer neutralised by the basic salts then it becomes acid and is inhabited by oxyphilous plants usually resulting in the formation of a **raised bog** which is a very acid peat. The commonest of these oxyphilous plants are species of bog moss, Sphagnum, which gives the alternative name of **moss peat.**

7. RAISED BOG

Raised bog usually develops on a fen as mentioned above, the surface of the fen having risen above the level of the ground water so that it is no longer neutralised and becomes inhabited by plants which can flourish in acid conditions. They may however be formed on top of a valley bog which is formed where water draining from acidic rocks, stagnates in a flat valley or in a depression and the soil is always wet.

The classification of soils according to the soil texture

As rock is weathered to form soil, so it is broken down into particles the majority being of one size range, which is characteristic for that particular rock, and which is referred to as the **soil texture**. Soils may be classified according to the soil texture. The sizes of the predominant or characteristic particles, which are named and classified by international agreement, are listed below:—

gravel particles	those which are larger than 2 mm diameter
coarse sand particles	those between 2 mm and 0.2 mm diameter
fine sand particles	those between 0.2 and 0.02 mm diameter
silt particles	those between 0.02 mm and 0.002 mm diameter
clay particles	those less than 0.002 mm diameter

1. GRAVEL

A soil made up mainly of particles with a diameter greater than 2 mm, mixed with particles of sand of less than 2 mm diameter and also some finer particles. Aeration is very good and percolation of water very free but the soil is poor in nutrients and is dry and generally unfavourable to plant life.

2. SANDY SOIL

In coarse sandy soil the size of the predominant particles varies between 2 mm and 0·2 mm diameter, in fine sand between 0·2 mm and 0·02 mm. Percolation is easy and aeration good, thus a sandy soil has a low water holding capacity and low capillarity, and is dry and light and is poor in nutrients due to leaching. The absence of bases causes a tendency towards acidity and it easily becomes podsolised.

3. SILT

Here the majority of particles are between 0·02 mm and 0·002 mm in diameter. This soil has a high water holding capacity and percolation, aeration and capillary rise of water are all fairly free. This is a soil favourable to plant growth, typically supporting meadowland or where wooded oak or alderwood.

4. CLAY

More than 30% of the particles have a diameter of less than 0·002 mm. In such a soil percolation of rain water is very slow and aeration poor. Clay soil has a very high water holding capacity and is therefore wet, heavy and cold being difficult to work. It also warms up slowly due to the high water content. Because of these characteristics it is unfavourable for the growth of many plants although it is rich in bases. It is improved by incorporation of humus.

5. LOAM

This is the most evenly balanced soil. Like the other soils it is composed of a mixture of particles of many sizes but unlike the other soils, without an overwhelming amount of any one type of particle. As a result it possesses most of the characteristics favourable to plant growth whilst lacking those which are unfavourable. The composition varies in different localities and one can have clay loams, medium loams and sandy loams.

Typical composition of loam:— 6-15% clay particles
40-60% silt particles
20-50% sand particles

Soils classified according to special chemical characteristics

1. CALCAREOUS SOILS

These soils contain an excess of calcium carbonate as in the climatic soil called a rendzina (described on page 130). They are formed from chalk or other limestones. The soil has a typical alkaline reaction and is inhabited by a specialised characteristic vegetation.

2. SALINE SOILS

Saline soils are those containing an excess of sodium chloride (common salt) and similar salts. They are mostly maritime, being formed in tidal estuaries where they may be flooded by the sea at high tide. Saline soils may also be found inland near salt springs. They support a specialised halophytic vegetation.

3. ORGANIC SOILS

These are the peats (see page 130). They are pure organic soils formed under conditions which prevent the decay of dead plant remains which therefore accumulate forming the acid organic peat.

4. LOCAL CHARACTERISTICS

Soils show various local characteristics, for example in Devonshire and some parts of Somerset the soil is red due to the large amount of iron salts in the soil, which is formed from red sandstone. In Kent and Sussex the soil is white and limey due to the presence of chalk.

SOIL STUDY EXPERIMENTS

Experiment 126
Mechanical analysis of soil, Method 1.   Using a series of graded sieves

APPARATUS REQUIRED
set of interlocking sieves which stand one above the other, the floors of the sieves being formed from wire mesh the size of which decreases from top to bottom of the set
sample of soil

METHOD OF PROCEDURE
Take the sieves and arrange them in series one above the other according to the size of the perforations in the wire mesh, the sieve with the perforations of largest diameter being at the top and the sieve with the smallest perforations being at the bottom.   Take a sample of the soil, and after weighing it accurately, place it in the top sieve which should then be covered with a lid.   Make sure that the lid is secure and that the sieves are firmly interlocked, then take the complete set and, holding them firmly, shake them vigorously for about 10 minutes. After shaking, separate the sieves and note whether all the soil still remains in the top sieve or whether it is distributed throughout the set, then remove and weigh separately the samples of soil remaining in each of the sieves.

The weight of soil in each of the sieves may be expressed as a percentage of the weight of the original sample of soil.   From these results the composition of the soil and the size of the predominant particles may be determined and the type of soil used identified.

Experiment 127
Mechanical analysis of soil, Method 2.   Using water

APPARATUS REQUIRED
2 measuring cylinders
sample of soil
water

METHOD OF PROCEDURE
Place 50 ml of water in one of the cylinders and 50 ml of soil in the other, then pour the water into the cylinder containing the soil, cover the open end of the cylinder which now contains the soil and water mixture and shake it vigorously, then leave the 'mixture' to settle.   After 1 hour note any layering which has taken place as the soil and water mixture has settled.   Also note the composition and thickness of each layer, and any decrease in the total volume of the soil and water mixture.

Repeat the experiment using samples of different types of soils noting any variations in the layering and any differences in the final total volumes.

Experiment 128
To demonstrate the flocculation of clay

APPARATUS REQUIRED
3 test tubes in stand
powdered china clay
water containing a trace of ammonia
2% potassium chloride solution
0·2% calcium chloride solution
decinormal hydrochloric acid

METHOD OF PROCEDURE
Make a fairly strong suspension of clay (about 2%), in water containing a slight trace of ammonia.   Place some of the clay suspension in each of the test tubes.   Add 1 ml of 2% potassium chloride to one test tube, 1 ml of 0·2% calcium chloride to the second, and 1 ml of decinormal hydrochloric acid to the third.   Note the time taken for the clay suspension to settle in each test tube.

Note: The clay suspension contains clay particles which are colloidal, carrying a negative charge.   If an excess of positively charged ions is added to the clay suspension then the negative charges on the clay particles will be partly or completely neutralised and the force holding the clay particles apart will be removed with the result that they flocculate to form aggregate particles.

Experiment 129

To determine the volume of air in a sample of soil

APPARATUS REQUIRED
2 measuring cylinders
sample of soil
water

METHOD OF PROCEDURE
Place 50 ml of water in one cylinder and 50 ml of soil in the other then pour the water into the cylinder containing the soil, cover the end of the cylinder now containing the soil and water mixture and shake it vigorously. After leaving them to settle for a few minutes note the total volume of soil and water in the cylinder and calculate the volume of air which was present in the soil.

This experiment can be repeated using samples of several different types of soils and a comparison of air content made.

Experiment 130

To determine the amount of water in a sample of soil

APPARATUS REQUIRED
sample of soil (air dried)
small porcelain basin
oven
desiccator
balance

METHOD OF PROCEDURE
First weigh the porcelain basin empty then place approximately 10 g of soil in it and re-weigh to determine accurately the weight of the sample of soil used. Place the basin with the soil in an oven, the temperature of which should be about 100°C and leave it for about 2 hours. Remove the basin and soil from the oven and place them in a desiccator leaving them to cool, then re-weigh them and return them to the oven for about 30 minutes. After this they should be cooled in the desiccator as before and re-weighed. This heating, cooling and weighing procedure should be repeated until the weight is constant.

From your results calculate the weight of the water in the sample of soil.

Note: The dried sample of soil may now be used in the experiment to determine the quantity of humus in a sample of soil, (see following experiment).

Experiment 131

To determine the weight of humus in a sample of soil

APPARATUS REQUIRED
sample of soil in an evaporating basin which has been dried and weighed as in experiment 130 above
oven
tripod and bunsen burner
sand tray or gauze
desiccator
balance

METHOD OF PROCEDURE
Take the sample of soil in the evaporating basin, the soil and basin having been repeatedly dried and weighed until of constant weight, (as in experiment 130 above); place them on the sand tray (or wire gauze) on a tripod over a large bunsen burner. Ignite the soil sample and continue to heat for several minutes then, after cooling the basin and soil in a desiccator, re-weigh.

From your results determine the percentage humus in the sample of soil.

Note: The loss of weight is due mainly to the burning of any organic matter (humus) present in the soil sample, although carbon dioxide is lost from any carbonates present and any carbon is also lost by burning.

Experiment 132

Simple experiment to indicate the presence of calcium carbonate in a soil sample

APPARATUS REQUIRED
2 test tubes
a rubber bung to fit one of the test tubes, the bung having a single hole through it
a piece of bent glass tubing to fit the hole in the rubber bung
sample of soil
dilute hydrochloric acid
limewater

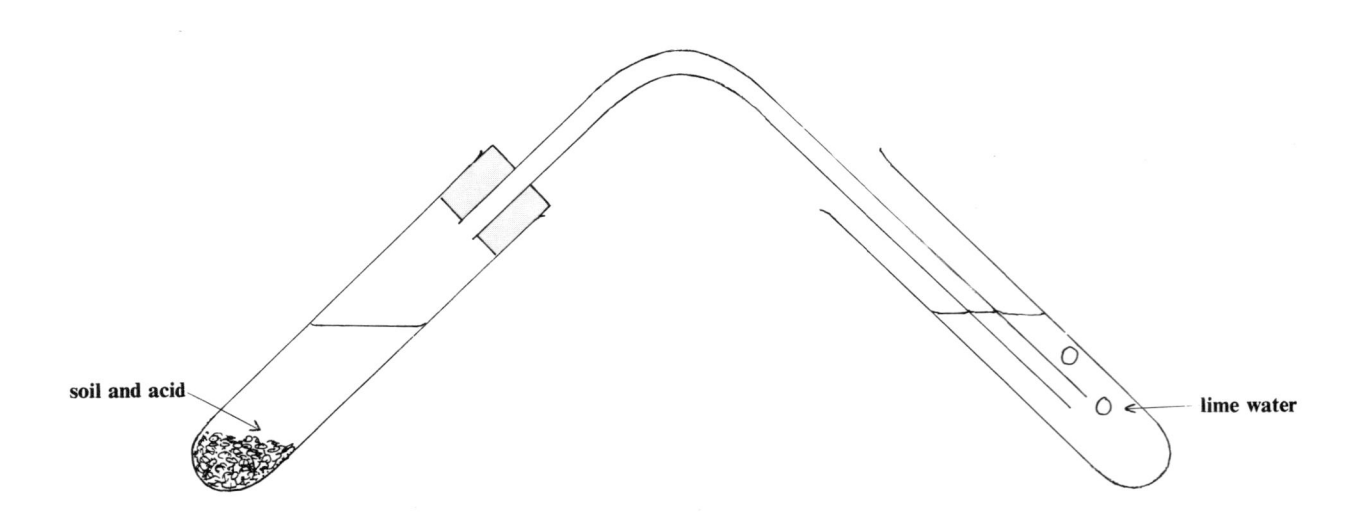

METHOD OF PROCEDURE
Fit the shorter limb of the piece of bent glass tubing through the hole in the rubber bung. Place some clear limewater in one of the test tubes and some soil in the other. Add some dilute hydrochloric acid to the soil in the test tube then quickly fit the rubber bung with the piece of the glass tubing into the neck of the test tube. Place the open end of the glass tubing below the surface of the limewater in the other test tube so that any gas escaping from the test tube containing the soil and acid mixture passes through the limewater. Note whether any gas is formed and any effect it may have on the limewater, and from this determine whether any carbonate was present in the sample of soil.

Note: Calcium carbonate reacts with dilute hydrochloric acid producing carbon dioxide. Carbon dioxide then reacts with limewater, which is a solution of calcium hydroxide, producing a fine white precipitate of calcium carbonate.

Experiment 133

To determine the acidity of a sample of soil

As explained in the text on page 128 the acidity of the soil may be determined using standard indicators with a known range of colour changes for known pH values. The indicators may be in liquid form or in the form of indicator papers. The most accurate measurements are of course made with indicator solutions which act over a fairly short pH range with distinct colour changes.

Soil samples should be taken from about 2 or 3 inches under the soil surface. A teaspoon or small trowel may be used for lifting them, and care should be taken not to touch them with your hands.

There are various soil testing kits available some of which may also be used for testing for the presence or absence of the various minerals essential for plant growth. Since the various kits are accompanied by full instructions for their use it is pointless to repeat them here.

Experiment 134

To demonstrate the presence of micro-organisms in the soil.   Method 1

APPARATUS REQUIRED
sample of soil
2 conical flasks
muslin
cotton
caustic soda
2 rubber bungs to fit the conical flasks, each with a single hole
2 pieces of bent glass tubing
2 glass dishes containing water
beaker

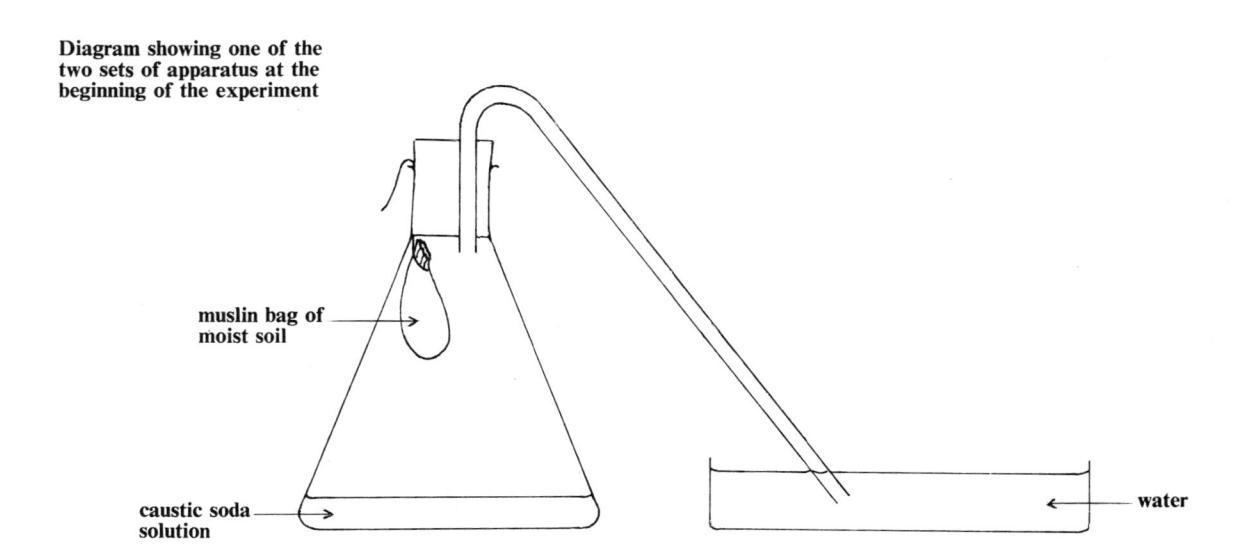

Diagram showing one of the two sets of apparatus at the beginning of the experiment

- muslin bag of moist soil
- caustic soda solution
- water

METHOD OF PROCEDURE

Take some of the soil, place it in the piece of muslin, form the muslin into a bag around the soil and tie securely with a piece of cotton. Take a second portion of soil, heat it to kill any micro-organisms present then after cooling tie this also in a piece of muslin. Moisten both bags of soil with water. Place some caustic soda in the bottom of each of the flasks. After fitting a piece of bent glass tubing in the hole, in each of the bungs, suspend the bags of soil one in each of the flasks, making sure that they are clear of the caustic soda solution and hold them in position by inserting a bung in the neck of each flask. Place the open end of each of the pieces of bent glass tubing in a dish of water then leave the apparatus for a few days. Note any change in the level of the liquids in the glass tubes and determine whether there are any micro-organisms in the samples of soil.

Note: All living organisms respire using up oxygen and giving out carbon dioxide. Any carbon dioxide produced in the flasks is absorbed by the caustic soda solution.

Experiment 135

To demonstrate the presence of micro-organisms in the soil.   Method 2.

APPARATUS REQUIRED
sample of soil
funnel
flask
industrial methylated spirits or 95% alcohol
source of bright light
glass wool
dark paper

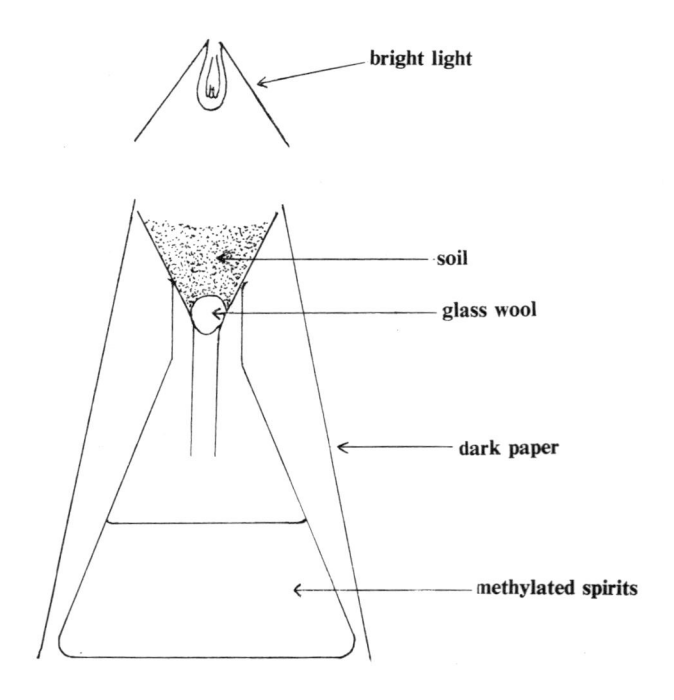

Diagram: Funnel containing soil with glass wool plug, inserted into a flask of methylated spirits, surrounded by dark paper, with a bright light above.

Labels: bright light, soil, glass wool, dark paper, methylated spirits

METHOD OF PROCEDURE

Place some of the methylated spirits in the flask, then place the funnel in the neck of the flask making sure that the bottom of the stem is clear of the liquid. Place a loose plug of glass wool in the top of the stem of the funnel then fill the funnel with soil. Surround the flask and funnel with a piece of dark paper, arrange a bright light directly over the funnel as shown in the diagram and leave for about 24 hours. Remove the paper from around the flask and identify any micro-organisms which have collected in the liquid in the flask.

Experiment 136

To demonstrate the variations in permeability of different soils and their ability to retain water

APPARATUS REQUIRED

samples of different types of soils such as clay, coarse and fine sand, loam, chalky soil etc.
sieve with 1mm perforations
several sets of the following, (one for each sample of soil):—
    piece of 1 inch diameter glass tubing about 2 ft long
    rubber bung (1 inch diameter to fit glass tubing) with a single hole through it
    piece of narrow bore glass tubing to fit hole in bung
    beakers
    measuring cylinder
    piece of muslin
    clamp and stand

METHOD OF PROCEDURE

Air-dry the samples of soil then sieve each sample using a sieve with 1 mm diameter perforations. Fit short pieces of narrow glass tubing into one end of each of the bungs, covering the other end of each bung with a piece of muslin. Fit each piece of wide bore tubing with a bung, muslin covered side innermost, then fill each tube with a different sample of soil, adding the soils slowly and tapping the tubes at the same time to ensure even settling of the soil particles. Add the soils until the tubes are about three quarters full making sure that they are all filled to the same level. Support each tube with a clamp and stand.

Place a measuring cylinder beneath the open end of each of the pieces of narrow glass tubing. Place 75 ml of water in each of the beakers, then in quick succession, pour the water from the beakers into the open ends of the soil filled tubes and leave.

Note the time taken in each case for the water to reach the bottom of the column and the first drop to fall into the cylinder. Then after 24 hours compare the amount of water in each of the cylinders.

Experiment 137

To demonstrate the capillary action of different types of soils

APPARATUS REQUIRED

samples of several different types of soils such as, clay, coarse and fine sand, loam, chalky soil etc.
sieve with 1mm perforations
several sets of the following (one for each soil sample):—
    piece of 1 inch diameter glass tubing about 2 ft. long
    glass dish
    piece of linen
    cotton thread
    clamp and stand
    water

Diagram showing one of the soil filled tubes at the beginning of the experiment

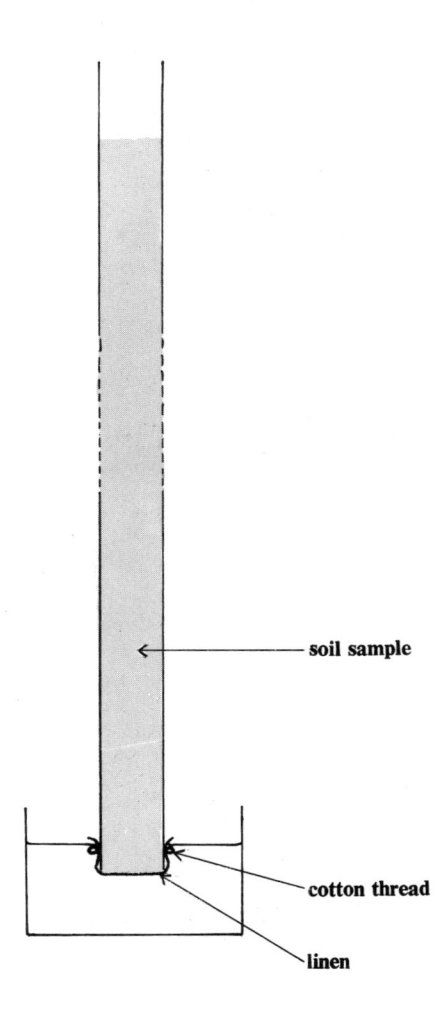

— soil sample

— cotton thread

— linen

METHOD OF PROCEDURE

Air-dry the soil samples then sieve each using a sieve with 1 mm perforations. Tie a piece of linen over one end of each piece of glass tubing, then take one of the tubes and one of the samples of soil, and fill the tube until it is about three quarters full, tapping the tube from time to time to make sure that the soil particles settle evenly. Fill each of the other tubes, with different samples of soil, in the same way. Then support each in a vertical position using a clamp and stand.

Place about 50 ml of water in each of the glass dishes then lower the linen covered end of each tube of soil into a dish of water. Compare the rates of rise of water upwards through the different soils, that is **compare the capillary action of the different soils**.

Subject Index

absorption
    active 6, 7.
    passive 6, 7.
adenosine diphosphate 31, 44.
adenosine triphosphate 31, 44.
adsorption chromatography 82-83.
aerobic respiration 45, 48, 49-51.
L-alanine 64.
aldehyde group 58.
α-amino acid 63, 64.
amino acids 63, 64.
    tests for 68.
amylopectin 60, 61
amylose 60.
anaerobic respiration 45, 47, 49.
apoplast 7.
α-L-arabinopyranose 63.
ascending paper chromatography 83, 87.
autonomic movements 111, 119.
    circumnutation 119, 125.
    cyclosis 119, 125.
    experiments 125.
auxin a (auxentriolic acid) 112.
auxin b (auxenolinic acid) 112.
auxin concentration . . . the effect on the growth of shoots and roots 101.
auxins
    herbicidal 103.
    naturally occurring 101.
    synthetic 103.
auxin theory 111, 112.
    and geotropic response of roots 112.
    and geotropic response of stems 113.
    and phototropic response of stems 113.

biological weathering (soils) 127.

Calvin's scheme 36, 37.
carbohydrates 58-63.
    tests for 66-68.
cellular respiration 44.
cellulose 61, 62.
    test for 67, 68.
chemical weathering (of rock) 126, 127.
chemonasty 115.
chemotaxis 117.
chemotropism 114.
cholesterol 65.
    test for 71.
chromatography 81-96.
    adsorption chromatography 82-83.
    ascending paper chromatography 83, 87.
    descending paper chromatography 83-87.
    distribution coefficient 83.
    experiments 89-96.
    ion exchange chromatography 88.
    general procedure 81.
    partition chromatography 83-88.
    preparation of sample for analysis 82.
    Rf value 87.
    thin layer chromatography 88.
circumnutation 119, 125.
citric acid cycle 48, 49.
climatic soils 129.
cuticular transpiration 9, 10.
cyclosis 119, 125.

dark stage of photosynthesis 34, 36, 37.
descending paper chromatography 83-87.
2.4.dichlorophenoxyacetic acid (2.4.D) 103.
diffusion pressure 1.

diffusion pressure deficit 1, 3.
disaccharides 59, 60.
    tests for 67.
distribution coefficient 83.

energy within the plant cell 31, 44.
enzymes 71-80.
    factors affecting enzymic reactions 72-75.
    enzyme experiments 75-80.

D-fructose 58.
D-fructofuranose 59.

α-D-galactopyranose 62.
gaseous exchange 44.
geotropism 112, 113.
gibberellic acid (A_3) 102.
gibberellins 102.
D-glucose 58.
α-D-glucopyranose 59.
β-D-glucopyranose 59.
glycine 63.
gross suction pressure 3, 4.
growth 97-110.
    experiments 103-110.
    growth pattern of a plant 97.
    plant growth hormones 99-103.

hemicelluloses 62.
herbicidal auxins 103.
heteroauxin (IAA) 101, 112.
humus 128.
hydrotropism 114.
hygroscopic movements 117-119.
hypertonic 2.
hypotonic 2.

incipient plasmolysis 4.
β-indolyl acetic acid (IAA) 101, 112.
inulin 61.
    test for 68.
ion exchange chromatography 88.
isotonic 2.

ketone group 58.
kinetin 102.
kinins 102.
Krebs' citric acid cycle 48, 49.

leaf suction 8.
lecithin 66.
    test for 71.
lenticular transpiration 9.
light stage of photosynthesis 34, 35.
lignin
    test for 68.
lipids 65, 66.
    tests for 71.

α-D-mannopyranose 62.
mechanism of photosynthesis 34.
meristems 98.
mineral uptake by plants 9.
mineral requirements of plants 8.
mitotic cell division 98-99.
    cytoplasmic division 99.
    nuclear division 98.
monosaccharides 58, 59.
    tests for 66, 67.

nastisms 114-116.
    chemonasty 115.
    nyctinastisms 115.
    seismonasty 116.
net suction pressure 3, 4.
nuclear division 98.
nyctinastisms 115.

oligosaccharides 59, 60.
osmosis 1-5.
    experiments 13-20.
    passage of water into and through the plant 6-8.
    special effects of solutes on plant cells 5.
    water relationships of plant cells 3-5.
osmotic pressure 2.

paratonic movements 111-119.
    hygroscopic movements 117-119.
    nastisms 114-116.
    tactic movements 117.
    tropisms 111-114.
    turgor movements 119.
partition chromatography 83-88.
pectic acid 63.
pectic compounds 63.
peptide bond 64.
peptides 64.
photolysis of water 34, 35.
photosynthesis 31-43.
    Calvins' scheme 36, 37.
    dark stage 34, 36, 37.
    experiments 37-43.
    light stage 34, 35.
    limiting factors 32.
    mechanism of 34.
    photolysis of water 34, 35.
    raw materials for 32.
photosynthetic pigments 32.
photosynthetic tissue 32.
phototaxis 117.
phototropism 113, 114.
plant growth hormones 99-103.
plant movement 111-125.
    autonomic movements 111, 119.
    experiments 120-125.
    paratonic movements 111-119.
plant water relationships 1-30.
    absorption 6, 7.
    osmosis 1-9.
    osmosis experiments 13-20.
    transpiration 9-12.
    transpiration experiments 20-30.
plasmolysis 4.
polypeptides 64.
polysacchardies 60-63.
    tests for 67, 68.
presentation time 111.
promeristem 98.
proteins 63-65.
    structure of 64, 65.
    tests for 69-70.

respiration 44-57.
    cellular (tissue) 44.
    aerobic 45, 48, 49-51.
    anaerobic 45-47, 49.
    experiments 53-57.
    factors affecting the rate of 52.
    Krebs' citric acid cycle 48, 49.
    respiratory chain (hydrogen transport system) 49-51.
    respiratory quotient 51, 52.
Rf value 87.

rock breaking 126.
rock rotting 126.
root pressure 7, 8.

sedentary soil 126.
seismonasty 116.
soil 126-137.
    air content 129.
    classification of soils 129-131.
    effect of climate on developing soil 127.
    experiments 132-137.
    formation 126.
    humus content 128.
    sedentary soil 126.
    soil acidity 128.
    soil profile 129.
    soil temperature 129.
    soil types 129-131.
    structure 128.
    transported soil 126.
    water content 129.
    weathering of rock 126, 127.
starch 60. tests for 67.

stomatal transpiration 9-11.
sucrose 60.
suction pressure 3, 4.
symplast 7.
synthetic auxins 103.

tactic movements 117.
threshold value 111.
thigmotropism 114.
thin layer chromatography 88.
tissue respiration 44.
translocation 9.
transpiration 9-12, 20-30.
    advantages of 10.
    cuticular 9-10.
    experiments 20-30.
    factors affecting the rate 10
    leaf suction 8.
    lenticular 9.
    plant adaptations to reduce the rate 12.
    stomatal 9-11.
transported soil 126.

tropsims 111-114.
    chemotropism 114.
    geotropism 112, 113.
    hydrotropism 114.
    phototropism 113, 114.
    thigmotropism 114.
    experiments to demonstrate tropisms 120.
tryptophan 64.
turgor movements 119.
turgor pressure 3.

wall pressure 3.
water potential 4.
water . . . entry into plant 6.
    entry into xylem 7.
    passage through plant 6.
    /plant cell relationships 3-5.
weathering of rock 126, 127.
    "climatic" weathering 126.
    chemical weathering 126, 127.
    biological weathering 127.
α-D-xylopyranose 63.

Index of Experiments

Expt. No. **OSMOSIS EXPERIMENTS**

1. Simple experiment to demonstrate osmosis 13
2. Microscopical demonstration of plasmolysis 14
3. To determine the diffusion pressure deficit (suction pressure) of cells from the epidermis of an onion bulb scale 14
4. To determine the diffusion pressure deficit (suction pressure) of the cells of a potato tuber 15
5. To determine the diffusion pressure deficit (suction pressure) of cells from a potato tuber by change in weight of discs of potato tissue 16
6. To determine the diffusion pressure deficit (suction pressure) of potato tissue by measurement of changes in the specific gravity of the external solution 16
7. To determine the effect of various treatments on the permeability of beetroot cells 17
8. To determine the effect of various solutes on cells from the epidermis of an onion bulb scale 17
9. To demonstrate the passage of water through plant tissue 18
10. To determine the effect of pH and temperature on the permeability of cells from beetroot 18
11. To determine the minerals essential for healthy plant growth using water culture solutions 19
12. To demonstrate tissue tensions using a dandelion peduncle 19
13. Experiment to demonstrate tissue tensions using a piece of Elder shoot 20

TRANSPIRATION EXPERIMENTS

14. To demonstrate that water vapour is given off by a leafy shoot 20
15. Simple experiment to demonstrate from which of the two surfaces of a leaf most water vapour is lost 21
16. To compare the rate of loss of water in the form of vapour, from the two surfaces of a leaf, using cobalt chloride paper 21
17. To compare stomatal and cuticular transpiration 22
18. To compare the stomatal densities of two leaves 23
19. Examination of stomata to determine their frequency per square centimetre of leaf surface 23
20. To demonstrate the use of a porometer 23
21. To determine the volume of the air space system of a leaf 24
22. To demonstrate transpiration by the loss of weight of a leafy shoot 25
23. To measure the rate of transpiration of a leafy shoot using a potometer 26
24. To compare the functioning of different types of potometer by calculation of the standard error of each 28
25. To demonstrate leaf suction 29
26. To demonstrate root pressure 30
27. To demonstrate the opening and closing mechanism of a stoma 30

PHOTOSYNTHESIS EXPERIMENTS

28. To demonstrate the presence of starch in a green leaf 37
29. To demonstrate that light is necessary for photosynthesis 38
30. To determine whether chlorophyll is necessary for photosynthesis 39
31. To demonstrate that carbon dioxide is necessary for photosynthesis 39
32. To determine whether a gas is given off as a result of photosynthesis and to identify it, also to determine whether photosynthesis occurs in the dark 40
33. To demonstrate chemically that oxygen is produced during photosynthesis (Kolkwitz' method) 41
34. To compare the rates of photosynthesis under differing conditions 41
35. To determine the effects of light and heat on the rate of photosynthesis 42
36. Ganong's disc method for determining the rate of photosynthesis 42
37. Extraction and separation of the photosynthetic pigments 43

Expt. No. **RESPIRATION EXPERIMENTS**

38. To demonstrate anaerobic respiration 53
39. To demonstrate that anaerobic respiration can occur in tissues normally respiring aerobically 54
40. To demonstrate the production of carbon dioxide as a result of aerobic respiration 54
41. To demonstrate the production of heat during respiration 55
42. To determine the respiratory quotient for barley grains 56
43. To demonstrate the effect of heat on the rate of respiration 57

BIOCHEMICAL TESTS

TESTS FOR CARBOHYDRATES 66
General tests 66
Tests for monosaccharides 66, 67
Tests for disaccharides 67
Tests for polysaccharides 67, 68
   (including test for lignin page 68)

TESTS FOR AMINO ACIDS 68
General reactions 68
Specific reactions 69

TESTS FOR PROTEINS 69
General tests 69

44. To determine the isoelectric point of protein 70

TESTS FOR LIPIDS 71

ENZYME EXPERIMENTS

45. To demonstrate the action of the enzyme phosphorylase 75
46. To demonstrate the action of the enzyme catalase 76
47. To demonstrate the localisation of the dehydrogenase enzymes with tetrazolium staining 76
48. To demonstrate the inhibition of polyphenol oxidase 77
49. To demonstrate the requirement of a prosthetic group by polyphenol oxidase 77
50. To demonstrate the requirement of a prosthetic group by catalase 78
51. To determine the pH optima of an enzyme using amylase 78
52. To determine the effect of temperature on an enzyme, using amylase 79
53. To demonstrate the action of the enzyme urease 79
54. To demonstrate, using the Thunberg technique, the presence of a dehydrogenase enzyme in milk 80

CHROMATOGRAPHIC EXPERIMENTS

55. Experiment to demonstrate the separation of amino acids by ascending paper chromatography 89
56. To demonstrate the separation of sugars by descending paper chromatography 90
57. To demonstrate the separation of the photosynthetic pigments by adsorption chromatography, using a column of icing sugar 91
58. To demonstrate the separation of the photosynthetic pigments by paper chromatography, using the ascending technique 93
59. To demonstrate the separation of the photosynthetic pigments by thin layer chromatography 93
60. To demonstrate the presence of sugars in plant tissues using descending paper chromatography 95
61. To demonstrate the presence of amino acids in plant extracts using ascending chromatography 96

Expt. No. **GROWTH EXPERIMENTS**

97 To determine whether water is necessary for germination 103
98 To determine whether air is necessary for germination 104
99 To examine the effect of temperature on germination 104
100 To determine by direct observation the growth rate of a root 104
101 To determine the growth rate of a root using a Nielson Jones auxonometer 105
102 To demonstrate the zones of growth in a root 106
103 To show the region of elongation in a shoot 107
104 To measure the rate of growth of a shoot using a smoked drum auxonometer 107
105 To measure the growth of a leaf 108
106 To demonstrate the effect of indolyle acetic acid (IAA) on the growth of oat coleoptiles 108
107 To demonstrate the effect of IAA on the growth of cress roots 109
108 To determine the effect of IAA on the rooting of plant cuttings 109
109 To show the effect of gibberellic acid on dwarf pea plants 110
110 To demonstrate the phytotoxic effects of 2.4.dichlorophenoxyacetic acid 110
111 To demonstrate the selective phytotoxic properties of 2.4.dichlorophenoxyacetic acid 110

EXPERIMENTS TO DEMONSTRATE TROPIC RESPONSES

Method of growing broad bean seedlings for use in experiments 120
Method used for marking the radicles of broad bean seedlings for use in experiments 120

112 To demonstrate the effect of gravity on a growing root 120
113 To demonstrate the influence of gravity on a growing shoot 121
114 To demonstrate that the geotropic response of a radicle is a definite growth movement and is not due to wilting 121
115 To demonstrate the effect on the growth of the radicles of broad bean seedlings of neutralising the unilateral influence of gravity 122

Expt. No. **Experiments to demonstrate tropic responses (contd.)**

116 To determine the presentation time of a radicle of a young bean seedling 123
117 To demonstrate the effect, on the growth of a young shoot, of neutralising the unilateral influence of gravity 123
118 To demonstrate hydrotropism in young bean seedlings 124
119 To determine which has more influence on the growth of the roots of young seedlings, either water of or gravity 124
120 To demonstrate thigmotropism 124

EXPERIMENTS TO DEMONSTRATE AUTONOMIC RESPONSES

121 Simple experiment to demonstrate cyclosis 125
122 To demonstrate circumnutation 125

SOIL STUDY EXPERIMENTS

126 Mechanical analysis of soil, Method 1. Using a series of graded sieves 132
127 Mechanical analysis of soil, Method 2. Using water 132
128 To demonstrate the flocculation of clay 132
129 To determine the volume of air in a sample of soil 133
130 To determine the amount of water in a sample of soil 133
131 To determine the weight of humus in a sample of soil 133
132 Simple experiment to indicate the presence of calcium carbonate in a soil sample 134
133 To determine the acidity of a sample of soil 134
134 To demonstrate the presence of micro-organisms in the soil. Method 1 135
135 To demonstrate the presence of micro-organisms in the soil. Method 2 135
136 To demonstrate the variations in permeability of different soils and their ability to retain water 136
137 To demonstrate the čapillary action of different types of soils 137

The Arlington Practical Botany

General Editor: Mary-Anne Burns, B.Sc.Hons.(Lond.), M.I.Biol.

Already Published

Book 1

PLANT ANATOMY

A.M.A. — 'Mrs. Burns has been eminently successful, and has produced a book of outstanding quality in this field. There is a minimum of text and yet there is as complete information as one could wish. The diagrams are excellent both in their function to instruct in plant anatomy and in the example they set of clarity, accuracy and simplicity. After a brief introduction to the student on drawing technique and preparation of materials, the author proceeds to present over 200 large, well-labelled diagrams covering all aspects of anatomy of the higher plants. In addition to the subject index there is a most useful index of plants used for illustrative purposes. One wonders how we managed to teach plant anatomy without this book and looks forward with anticipation to the succeeding volumes in the series.'

The Times Educational Supplement — 'The standard of this first book augurs well for the whole series, which should be warmly welcomed by both teachers and students . . . The book is extremely well produced. . . . No fewer than 230 admirable drawings are included, all clearly labelled, and with the magnification always precisely indicated. Concise theoretical explanations accompany the illustrations and these would be sufficient to make the book useful even for students who have little opportunity to pursue a full practical course. The opening pages, however, provide numerous practical hints about preparing sections, staining, maceration, preserving, etc., and the best ways in which to make drawings from specimens.'

School Science Review — 'An extremely useful and comprehensive guide to practical plant anatomy. The introduction offers sound advice on drawing, there are excellent indices to subjects and species, and between these are 230 drawings with detailed notes covering all aspects of flowering plant anatomy. . . . This book satisfies a very real need and the succeeding volumes will be awaited eagerly.'